# Quantum Science and Technology

For further volumes:
http://www.springer.com/series/10039

# Quantum Science and Technology

## Aims and Scope

The book series Quantum Science and Technology is dedicated to one of today's most active and rapidly expanding fields of research and development. In particular, the series will be a showcase for the growing number of experimental implementations and practical applications of quantum systems. These will include, but are not restricted to: quantum information processing, quantum computing, and quantum simulation; quantum communication and quantum cryptography; entanglement and other quantum resources; quantum interfaces and hybrid quantum systems; quantum memories and quantum repeaters; measurement-based quantum control and quantum feedback; quantum nanomechanics, quantum optomechanics and quantum transducers; quantum sensing and quantum metrology; as well as quantum effects in biology. Last but not least, the series will include books on the theoretical and mathematical questions relevant to designing and understanding these systems and devices, as well as foundational issues concerning the quantum phenomena themselves. Written and edited by leading experts, the treatments will be designed for graduate students and other researchers already working in, or intending to enter the field of quantum science and technology.

Kia Manouchehri · Jingbo Wang

# Physical Implementation
# of Quantum Walks

 Springer

Kia Manouchehri
Jingbo Wang
School of Physics
The University of Western Australia
Perth
Australia

ISBN 978-3-642-44709-9          ISBN 978-3-642-36014-5 (eBook)
DOI 10.1007/978-3-642-36014-5
Springer Heidelberg New York Dordrecht London

# Preface

Random walks have been employed in virtually every science related discipline to model everyday phenomena such as biochemical reaction pathways and DNA synapsis (Sessionsa et al. 1997), the genomic distance from DNA sequence location in cell nuclei (van den Engh et al. 1992), optimal search strategies for hidden targets such as animals' foraging (Bénichou et al. 2005), diffusion and mobility in materials (Trautt et al. 2006), the trail of a particle undergoing brownian motion (Stewart 2001) as well as exchange rate forecast (Kilian and Taylor 2003). They have also found algorithmic applications, for example, in solving differential equations (Hoshino and Ichida 1971), quantum monte carlo for solving the many body Schrödinger equation (Ceperley and Alder 1986), optimization (Berg 1993), clustering and classification (Schöll and Schöll-Paschingerb 2003), fractal theory (Anteneodo and Morgado 2007) or even estimating the relative sizes of Google, MSN and Yahoo search engines (Bar-Yossef and Gurevich 2006).

Whilst the so called *classical* random walks have been successfully utilized in such a diverse range of applications, *quantum* random walks are expected to provide us with a new paradigm for solving many practical problems more efficiently (Aharonov et al. 1993; Knight et al. 2003b). In fact quantum walks have already inspired efficient algorithms with applications in connectivity and graph theory (Kempe 2003b; Douglas and Wang 2008), as well as quantum search and element distinctness (Shenvi et al. 2003; Childs and Goldstone 2004b), due to their non-intuitive and markedly different properties, including faster *mixing* and *hitting* times. And more recently, some quantum walk processes are shown to be capable of acting as universal computational primitives (Childs 2009).

The emerging prospects for a new generation of quantum algorithms inspired by quantum walks have naturally fuelled a second area of research: developing the physical "hardware" that is capable of performing a quantum walk in the laboratory. As well as being experimentally viable, such a physical implementation is expected to provide a natural mechanism by which it can be scaled up, enabling it to deal with modestly large practical problems. Moreover, while purpose built systems for implementing specific quantum walk algorithms may be more straightforward to design, when considering a hypothetical problem formulated in terms of a particular type of graph, developing a problem-independent implementation scheme that is not limited to specific connectivity criteria is highly desirable.

Over the past decade there have been several proposals for implementing quantum walks, utilizing a variety of quantum, classical and hybrid systems including Nuclear Magnetic Resonance, cavity QED, ion traps, optical traps, optical networks and quantum dots. Of these, a vast majority provide constructive insights into the elements of a successful physical implementation, though in themselves are unsuitable as a practical scheme. Other proposals have considered the notions of feasibility, scalability and generality of application, but only to a limited extent. Nonetheless, while building a large scale quantum walk machine remains a considerable challenge in the foreseeable future, "proof of principle" implementations have already been experimentally demonstrated for a number of proposals.

We begin this book with a brief overview of quantum walk theory, including a description of the two main classes of walks, namely, *continuous-time* quantum walks and *discrete-time* or *coined* quantum walks, as well as their properties and applications; areas which have already received substantial treatment in other reviews including those of Kempe (2003b), Ambainis (2003), Kendon (2007), and Venegas-Andraca (2012). The main focus of this book however will be the *physical implementations* of quantum walks examined in the subsequent chapter, where we present a comprehensive survey of numerous implementation schemes to date. The tremendous diversity of approaches in these proposals has necessitated references to a wide array of underlying physical phenomena, particularly in relation to the field of quantum optics. Therefore, to assist the reader while maintaining continuity throughout the book, a considerable body of supplementary material and background theory has been included in the appendices.

In carrying out the original research described in this book, we have greatly benefited from valuable and stimulating discussions with many pioneering theorists and experimentalists in this field, in particular Gerard Milburn, Jason Twamley, Gavin Brennen, Peter Rohde, Jeremy O'Brien, Paolo Metaloni, Dieter Meschede, Jonathan Matthews, Andreas Schreiber, Norio Konno, Etsuo Segawa, Yutaka Shikano, Armando Perez, and Miklos Santha, for which we are truly grateful. We would also like to thank Zhijian Li, Michael Delanty, and Stefan Danilishin for their careful and critical proofreading of the manuscript, although any errors or omissions remain solely our responsibility. Special mention should be made of a number of students, most notably Brendan Douglas, Scott Berry, and Thomas Loke, who contributed towards our original research work presented here. The University of Western Australia has provided a rich intellectual environment that led to the completion of this book. The support and encouragement of Ian McArthur and Jim Williams at the School of Physics, as well as Angela Lahee (the editor) and Priya Balamurugan (the production editor) of Springer are also sincerely acknowledged.

Last but certainly not least, our deepest gratitude goes to our families. Kia thanks his wife, the wonderful Shameem, for her patient and never-ending encouragement of his aspirations. Jingbo thanks her ever-supportive husband Jie and her two children Edward and Eric, who have been a constant source of joy and inspiration.

# Contents

# Chapter 1
# Theoretical Framework

There are two broad classes of quantum walks, one in discrete-time and the other in continuous-time, each of which have independently emerged out of the study of unrelated physical problems. Here we present the theoretical background for both classes and briefly discuss some of their properties.

## 1.1 Discrete-Time Quantum Walks

### 1.1.1 Introductory Theory

Consider Alice at an intersection (Fig. 1.1), flipping a coin to decide whether to move to the left or right. We use the notation $+$ and $-$ to represent the result being heads or tails. Likewise $P_+$ and $P_-$ represent the probabilities for each outcome, where for an ordinary unbiased coin we expect $P_+ = P_- = 0.5$. Alice has decided to move to the right if she gets $+$ and left if she gets $-$. Before flipping the coin there is a $P_+$ chance that she would take the right path and a $P_-$ chance that she would take the left path. However after the coin flip there is no ambiguity and she makes the move to the right or left with absolute certainty. If Alice is originally at $x = 0$, then moving to the right or left represent stepping to position $x = +1$ or $x = -1$. She can continue using coin flips to move along a *decision tree* as illustrated in Fig. 1.2. This procedure is known as a (classical) random walk. It is well known that for such a random walk the final probability distribution is a Gaussian.

In a quantum analogue of the random walk, Bob uses a "quantum coin" with two "sides" which in the language of quantum mechanics are known as *basis states* and are conveniently represented by $|+\rangle$ and $|-\rangle$ using Dirac notation. Unlike the classical coin however, the quantum coin has the unintuitive property that as long as Bob does not look at it (i.e. does not perform an observation or measurement) it can be in a superposition of states. In other words if $|c\rangle$ represents the state of the

J. Wang and K. Manouchehri, *Physical Implementation of Quantum Walks*,
Quantum Science and Technology, DOI 10.1007/978-3-642-36014-5_1,
© Springer-Verlag Berlin Heidelberg 2014

**Fig. 1.1** Alice at an intersection resorts to a coin flip to decide which path to take

**Fig. 1.2** Alice makes her way through a decision tree using a *classical* random walk. The *red path* indicates Alice's decisions following each coin flip

quantum coin after being "flipped", it can simultaneously exist in both states such that

$$|c\rangle = \alpha_+|+\rangle + \alpha_-|-\rangle, \tag{1.1}$$

where $\alpha_+ = \langle +|c\rangle$ and $\alpha_- = \langle -|c\rangle$ are the complex amplitudes for the coin to be in each state. But if Bob looks at the coin in order to determine which way to move, he would then find the coin in either of the states (in the language of quantum mechanics the coin state has collapsed into either of the basis states) with a probabilities $P_+ = |\alpha_+|^2$ and $P_- = |\alpha_-|^2$. Hence if Bob was to determine the state of the quantum coin at every step, the final probability distribution would be identical to Alice's classical walk. Instead, while the quantum coin is in a superposition state, being a quantum walker himself, Bob simultaneously moves to the right *and* left with amplitudes $\alpha_+$ and $\alpha_-$ respectively. So starting at position

**Fig. 1.3** Bob makes his way through a decision tree using a discrete-time *quantum* random walk. The various shades of his figure are an impression of Bob's quantum wave-function as it spreads throughout the tree resulting in an interference pattern between the amplitudes traversing different paths

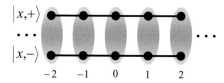

**Fig. 1.4** States of a discrete-time quantum walk on a line. The nodes or position states of the walk are labeled by $|x\rangle = \ldots, -2, -1, 0, 1, 2, \ldots$ and each node possesses two sub-nodes or coin states labeled by $|+\rangle$ and $|-\rangle$

state $|x = 0\rangle$, the move would put Bob in a superposition of position states $|+1\rangle$ and $|-1\rangle$. Similar to the situation with the quantum coin however, if we tried to find out where Bob actually was after the move, the act of observation would collapse Bob's position state and we would find him in either of the states with a probability $P_+ = |\alpha_+|^2$ and $P_- = |\alpha_-|^2$, again identical to the classical case. Hence the quantum walk proceeds in the absence of any observation or measurement during the walk, whereby Bob simultaneously arrives at multiple position and coin states with various amplitudes which constitute a *probability wave-function* depicted in Fig. 1.3.

Figure 1.4 illustrates the basic structure of the quantum states of the walk. The nodes along the horizontal axis represent the position states $|x\rangle$ of the quantum walker with $x = \ldots -1, 0, +1 \ldots$, while each node is comprised of two sub-nodes or sub-levels which represent coin states $|+\rangle$ and $|-\rangle$, also known as internal states, at that position. Hence the complete state of the Bob-coin system can be described as a superposition of all states given by

$$|\psi\rangle = \sum_x \alpha_{x,-}|x, -\rangle + \alpha_{x,+}|x, +\rangle, \tag{1.2}$$

where $\alpha_{x,+} = \langle x, +|\psi\rangle$ and $\alpha_{x,-} = \langle x, -|\psi\rangle$ are the complex amplitudes for the coin to be in states $|+\rangle$ and $|-\rangle$ respectively, while Bob is in the position state $|x\rangle$, and we have

$$\sum_x |\alpha_{x,+}|^2 + |\alpha_{x,-}|^2 = 1. \tag{1.3}$$

In this all-quantum system a "coin flip" takes a new meaning. Unlike the classical random walk in which flipping the coin produces a single state (head or tail) with probability $P_+$ or $P_-$, in the quantum case it corresponds to a unitary rotation of the coin basis states given by

$$\begin{pmatrix} \alpha'_+ \\ \alpha'_- \end{pmatrix} = \hat{C} \begin{pmatrix} \alpha_+ \\ \alpha_- \end{pmatrix}, \tag{1.4}$$

where $(\alpha_+ \ \alpha_-)^T$ and $(\alpha'_+ \ \alpha'_-)^T$ are the vector representations of the coin states $|c\rangle$ and $|c'\rangle$ before and after the coin flip, and the *coin operator*

$$\hat{C} = \begin{pmatrix} \cos(\theta) & e^{i\phi_1}\sin(\theta) \\ e^{i\phi_2}\sin(\theta) & -e^{i(\phi_1+\phi_2)}\cos(\theta) \end{pmatrix} \tag{1.5}$$

is an arbitrary $2 \times 2$ unitary matrix up to a global phase. This change of basis can be equally represented via the mapping

$$|+\rangle \longrightarrow \cos(\theta)|+\rangle + e^{i\phi_1}\sin(\theta)|-\rangle, \quad \text{and} \tag{1.6}$$

$$|-\rangle \longrightarrow e^{i\phi_2}\sin(\theta)|+\rangle - e^{i(\phi_1+\phi_2)}\cos(\theta)|-\rangle. \tag{1.7}$$

The simultaneous stepping of the quantum walker to the left and right according to the state of the coin is formally achieved by acting on the system using a unitary *conditional translation operator* $\hat{T}$ which has the property that

$$\hat{T}|x, +\rangle \equiv \hat{T}_+|x, +\rangle = |x + 1, +\rangle \tag{1.8}$$

$$\hat{T}|x, -\rangle \equiv \hat{T}_-|x, -\rangle = |x - 1, -\rangle. \tag{1.9}$$

With the system initially in state $|\psi_0\rangle$, a single step in the quantum walk involves the simultaneous application of the coin operator $\hat{C}$, followed by the conditional translation operator $\hat{T}$, across all the states as depicted in Fig. 1.5. Figure 1.6 illustrates the first two steps of the walk for the initial state

$$|\psi_0\rangle = \frac{1}{\sqrt{2}} (|0, +\rangle + |0, -\rangle), \tag{1.10}$$

and the balanced coin

$$\hat{C} = \frac{1}{\sqrt{2}} \begin{pmatrix} 1 & i \\ i & 1 \end{pmatrix}. \tag{1.11}$$

After $n$ steps of the walk, the probability distribution for the quantum walk is simply given by

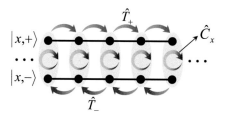

**Fig. 1.5** The operators responsible for driving a discrete-time quantum walk. Each step of the walk involves the action of local coin operators $\hat{C}_x$ responsible for mixing the coin states of individual nodes (*circular arrows*), followed by the action of the conditional translation operators $\hat{T}_+$ and $\hat{T}_-$ responsible for shifting the coin states (*left* and *right arrows*)

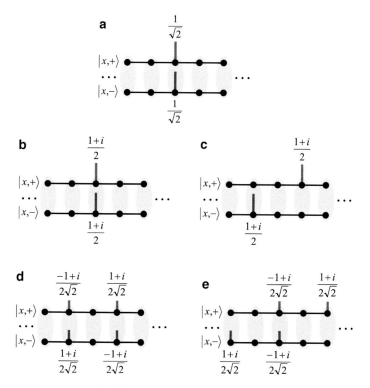

**Fig. 1.6** The first two steps of a discrete-time quantum walk. (**a**) The initial wave-function $|\psi_0\rangle = 1/\sqrt{2}(|0,+\rangle + |0,-\rangle)$. (**b**) and (**d**) The state of the walk following the application of local coin operators given by Eq. 1.11. (**c**) and (**e**) The state of the walk following the application of the translation operator $\hat{T}$ given by Eqs. 1.8 and 1.9

$$P_n(x) = |\langle x, -|\psi_n\rangle|^2 + |\langle x, +|\psi_n\rangle|^2. \qquad (1.12)$$

In other words, making a measurement on the position of the walker would collapse its wave-function into a single position state $|x\rangle$ with probability $P_n(x)$. This of

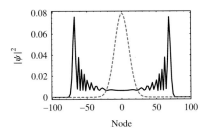

**Fig. 1.7** A comparison between the probability distributions of a classical (*dotted line*) and a discrete-time quantum (*solid line*) random walk and on a line after 100 steps. The quantum walk was propagated using an initial wave-function $|\psi_0\rangle = 1/\sqrt{2}(|0,+\rangle + |0,-\rangle)$ and identical local coin operators given by Eq. 1.11

course means that experimentally one has to repeat the walk procedure many times and measure the outcome in order to build up the probability distribution.

Figure 1.7 shows the probability distribution for the above quantum walk after $n = 100$ steps. A comparison with the classical random walk distribution, on the same plot, immediately reveals the rather surprising spreading properties of the quantum walk. For the classical walk the probability to find Alice after $n$ steps is a Gaussian distribution with a variance $\sigma^2$ growing linearly with time, so the expected distance from the origin is of the order $\sigma \sim \sqrt{n}$. By contrast the variance of the quantum walk scales as $\sigma^2 \sim n^2$, from which it follows that the expected distance from the origin is of the order $\sigma \sim n$. In other words the quantum walk is a ballistic process propagating quadratically faster than the classical walk which is a diffusive process (Ambainis et al. 2001; Blanchard and Hongler 2004).

What is significant here is that this non-classical distribution is due to the interference of the complex amplitudes as the quantum walk evolves and relies purely on the wave nature of the quantum walker (Knight et al. 2003a,b). As we will see later, understanding the quantum walks as an interference phenomenon has profound implications for their physical implementation.

### 1.1.2  The Pioneering Model

At this point it is instructive to briefly discuss the original work of Aharonov et al. (1993) in which they coined the term "quantum random walk" for the first time. There the authors introduced the idea of a random walk using a quantum coin whose basis states could be rotated after each steps. Their description of the quantum walk however was different from the above contemporary approach to quantum walks in two fundamental ways.

The first difference pertains to the relationship between the position states of the walk and the width of the walker. In the contemporary quantum walk theory, the position states are always assumed to form an orthogonal basis for the walker, that is $\langle x_j | x_i \rangle = \delta_{ij}$. In other words, after measuring the position of the walker, if it was found to be in state $|x_i\rangle$ with a probability equal to 1, then the probability

**Fig. 1.8** A discrete-time quantum walk on a continuous position space. The walker has a finite width $\Delta x$ centered on the marked nodes or position states of the walk which are a distance $\ell$ apart

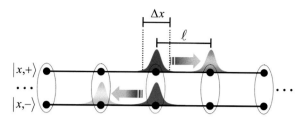

to find the walker in any other states $|x_j\rangle$ will be zero. This of course is trivially obvious in Fig. 1.6, where the position states are represented as discrete nodes $x = \ldots, -1, 0, +1, \ldots$ with an infinitely narrow width. In this case the nodes can be made arbitrarily close without affecting the outcome. The situation changes however when variable $x$ is continuous and the quantum walker has a finite width, e.g. represented by a Gaussian wave-packet with spatial width $\Delta x$. In this case to maintain orthogonality of states it becomes necessary to ensure that the step length $\ell$ between the nodes or position states of the walk obeys $\ell \gg \Delta x$ and the translation operator $\hat{T}$ displaces the walker from $x$ to $x \pm \ell$. Hence as illustrated in Fig. 1.8 the overlap between the distributions in neighboring sites becomes negligible and the amplitude $\langle x, c | \psi \rangle$ corresponds to the area under the curve which is in general complex valued. Although this point is not explicitly addressed in much of the literature, it is implicit in many of the physical implementations described in this book, including the work of Sanders and Bartlett (2003) in which they considered well separated coherent states of radiation with a finite width in phase space, as well as implementations based on optical lattices and optical microtraps where atoms are trapped in a periodic optical potential with lattice periods greater than the width of the atomic wavefunction. In their original proposal however Aharonov et al. (1993) did not require such orthogonality of position states. Instead in order to study the properties of the quantum walk analytically by making certain approximations, they assumed the opposite case, where $\ell \ll \Delta x$, i.e. the spatial width of the state is much larger than $\ell$.

In a related work, Aslangul (2005) described a thought experiment where a two level particle, represented by spin states $|+\rangle$ and $|-\rangle$, exhibits spin dependent velocity (i.e. the particle moves to the left or right depending on its spin state). A tuneable parameter $\omega$ in the model allows one to control the particle's spin flip rate, i.e. the frequency at which the spin basis is rotated by $\hat{C}$. The particle with wave-packet width $\Delta x$ and velocity $v$ would then hop between sites that are a distance $\ell = v/\omega \gg \Delta x$ apart. For small values of $\omega$ the particle moves a distance appreciably larger than $\Delta x$ before the spin is flipped and the direction of motion is changed. This represents a coined quantum walk on a discrete position space. Increasing $\omega$ however would result in a shorter travel distance $\ell$ between the spin flips, which gradually becomes comparable with $\Delta x$ leading to a smearing of the well-defined hopping sites and the position states are no longer orthogonal.

Another major difference between the contemporary quantum walk and the model described by Aharonov et al. (1993) is their measurement scheme. In our earlier description of the quantum walk we emphasized that the walk is allowed to

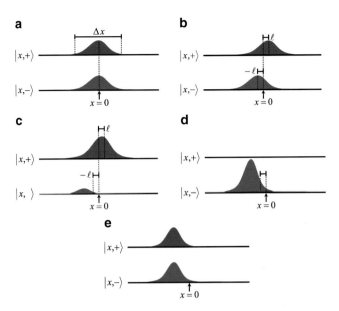

**Fig. 1.9** One step of a discrete-time quantum walk evolution for a walker with a finite-width $\Delta x$. (a) The initial wave-function $|\psi_0\rangle = 1/\sqrt{2}(|0,+\rangle + i|0,-\rangle) \otimes \mathcal{G}(x)$, where $\mathcal{G}$ is a Gaussian distribution. (b) The application of the conditional translation operator shifts the coin states by a distance $\ell \ll \Delta x$. (c) The application of a unitary coin operator mixes the coin amplitudes. (d) The (less likely) outcome of a coin state measurement which finds the walker to be in state $|-\rangle$; the quantum walker's position marked by the center of its wave-packet has been shifted beyond the classically expected range $\ell$. (e) The coin state of the walk is reset in preparation for the next step

proceed for many steps and the state of the walker is only measured once at the end of the walk. Aharonov et al. (1993) however suggested measuring the state of the quantum coin after each step but without making any such measurements on the position state of the walker. Significantly however they were still able to demonstrate faster than classical spreading properties of the quantum particle when the coin was made unbalanced (i.e. $\theta \neq \pi/4$ in Eq. 1.5). This is illustrated in Fig. 1.9. Starting with the walker represented by a Gaussian distribution with spatial width $\Delta x$, initially in a balanced superposition of coin states centered at $x = 0$ (Eq. 1.10), the authors used a rotation matrix $\hat{C}$ (Eq. 1.5) with $\theta = \tan^{-1}(1 + \epsilon)$ and assumed $\ell/\Delta x \ll |\epsilon| \ll 1$ (Fig. 1.9a). Under these conditions they applied the translation operator $\hat{T}$ (Fig. 1.9b) followed by the coin operator $\hat{C}$ (Fig. 1.9c), before measuring the state of the coin in the rotated basis. It is easy to see that because of the unbalanced rotation the walker is now more likely to be found in one of the coin states than the other. Also because the position state remains unmeasured, each coin state measurement outcome corresponds to a normalized spatial distribution with its center shifted to the left or right. What is interesting however is the counter-intuitive outcome that even though the translation operator displaces only by $\ell$, for the less likely measurement outcome, the walker (i.e. the mean position

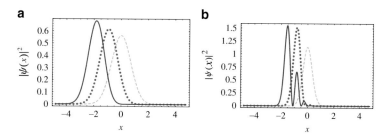

**Fig. 1.10** Distribution of $|\psi(x)|^2$ in Fig. 1.9 after $n = 0$ (*dashed*), 5 (*dotted*) and 10 (*solid*) steps, if one detects, successively, the walker to be in state $|-\rangle$ which involves a very small probability. In both plots the coin operator in Eq. 1.5 is parameterized by $\theta = \tan^{-1}(1 + \epsilon)$ with $\epsilon = -0.1$, the translation operator shift distance $\ell = 0.01$ and the initial Gaussian $\mathcal{G}(x) = \exp(-x^2/2\sigma^2)/(\pi\sigma^2)^{1/4}$, where $\sigma = 1$ in (**a**) and $\sigma = 0.5$ in (**b**)

of the spatial distribution) is shifted by $\delta x > \ell$ (Fig. 1.9d). In other words, the displacement of the walker is not bounded by the classically possible displacement. The walk can be extended to multiple steps by re-initializing the state of the coin to a balanced superposition state (Fig. 1.9e) after each measurement and repeating the above procedure. Using an initial Gaussian given by $\exp(-x^2/2\sigma^2)/(\pi\sigma^2)^{1/4}$ and for $\sigma = 1$, $\ell = 0.01$ and $\epsilon = -0.1$, Fig. 1.10a demonstrates the resulting spatial distribution after $n = 0, 5$ and 10 steps, if one detects, successively, the walker to be in state $|-\rangle$ which involves a very small probability. The plot displays a small deformation which becomes more relevant as the number of steps increases. Figure 1.10b demonstrates the large deformation if the initial Gaussian had a narrower width, using $\sigma = 0.5$. Aharonov et al. (1993) also proposed a physical implementation of this original model which was later improved by Agarwal and Pathak (2005) (Sect. 3.3).

It should be pointed out here that the mechanics of what is now referred to as a discrete-time quantum walk was in fact discovered much earlier in connection with the Dirac equation. As discussed by Meyer (1996), this goes back to Feynman's "checkerboard", a discrete space-time path integral that, in the continuum limit, generates the propagator for the Dirac equation in 1D. It was not until Aharonov et al. (1993) independently rediscovered the quantum walk however, that it became the subject of extensive analysis particularly in the context of quantum computation.

### 1.1.3   A More Rigorous Discussion

What we have described so far is the simplest form of quantum walks, namely the discrete-time quantum walk on an infinite line. In this section we will present a more comprehensive mathematical framework which underpins the extension of this basic model to higher dimensions and indeed to graphs with arbitrary complexity.

We begin by considering a position Hilbert space $\mathcal{H}_x$ comprised of basis states $\{|x\rangle\}_{x=-\mathcal{N}/2}^{\mathcal{N}/2-1}$, where $\mathcal{N}$ is the number of nodes traversed by the walker along a line in 1D, and a coin Hilbert space $\mathcal{H}_c$ comprised of basis states $\{|c\rangle\}_-^+$. Then the Hilbert space of the quantum walk $\mathcal{H}_w = \mathcal{H}_x \otimes \mathcal{H}_c$ consisting of basis states $\{|x\rangle \otimes |c\rangle\} \equiv \{|x, c\rangle\}$ (with the upper and lower bounds implicit). The quantum state of the walker can therefore be written as a linear superposition of these basis states

$$|\psi\rangle = \sum_{x=-\mathcal{N}/2}^{\mathcal{N}/2-1} \sum_{c=-}^{+} \alpha_{xc} |x, c\rangle \tag{1.13}$$

or equivalently represented by a $2\mathcal{N}$ dimensional vector

$$|\psi\rangle = \begin{pmatrix} \vdots \\ \alpha_{-1,-} \\ \alpha_{-1,+} \\ \alpha_{0,-} \\ \alpha_{0,+} \\ \alpha_{1,-} \\ \alpha_{1,+} \\ \vdots \end{pmatrix}. \tag{1.14}$$

A *global* coin operator $\hat{C}$ is then defined by an $\mathcal{N} \times \mathcal{N}$ block diagonal matrix

$$\hat{C} = \begin{pmatrix} \ddots & 0 & 0 & 0 & 0 \\ 0 & \hat{C}_{-1} & 0 & 0 & 0 \\ 0 & 0 & \hat{C}_0 & 0 & 0 \\ 0 & 0 & 0 & \hat{C}_1 & 0 \\ 0 & 0 & 0 & 0 & \ddots \end{pmatrix}, \tag{1.15}$$

where each $2 \times 2$ block $\hat{C}_x$ is a *local* coin operator given by Eq. 1.5, acting on the coin states of node $x$ such that

$$\hat{C}|\psi\rangle = \sum_{x=-\mathcal{N}/2}^{\mathcal{N}/2-1} \hat{C}_x \sum_{c=-}^{+} \alpha_{xc} |x, c\rangle$$

$$= \sum_{x=-\mathcal{N}/2}^{\mathcal{N}/2-1} \sum_{c=-}^{+} \alpha'_{xc} |x, c\rangle. \tag{1.16}$$

The action of the local coin operator on a single state of the walk can be equivalently represented by the mapping

$$|x, \pm\rangle \xrightarrow{\hat{C}_x} \gamma_{+,\pm}|x, +\rangle + \gamma_{-,\pm}|x, -\rangle, \tag{1.17}$$

where $\gamma_{+,\pm}$ and $\gamma_{-,\pm}$ are the column entries in $\hat{C}_x$. It is often desirable however to design the global coin operator such that identical local coins are applied to each node. In other words by letting $\hat{C}_{-\mathcal{N}/2} = \ldots = \hat{C}_{\mathcal{N}/2-1} = \hat{C}$ the global coin operator can be written in the separable form

$$\hat{\mathcal{C}} = \hat{I} \otimes \hat{C}, \tag{1.18}$$

where the identity operator $\hat{I}$ is $\mathcal{N}$-dimensional (Shenvi et al. 2003). A particularly common example of this is the so called Hadamard walk, where all the local coins

$$\hat{C} = \frac{1}{\sqrt{2}} \begin{pmatrix} 1 & 1 \\ 1 & -1 \end{pmatrix} \tag{1.19}$$

are the Hadamard matrix. The action of the conditional translation operator can be formally defined as

$$\hat{\mathcal{T}}|\psi\rangle = \left( \hat{T} \otimes \hat{P}_+ + \hat{T}^\dagger \otimes \hat{P}_- \right) |\psi\rangle$$

$$= \sum_{x=-\mathcal{N}/2}^{\mathcal{N}/2-1} \left( \hat{T}|x\rangle \otimes |+\rangle\langle+| \sum_{c=-}^{+} \alpha_{xc}|c\rangle + \hat{T}^\dagger|x\rangle \otimes |-\rangle\langle-| \sum_{c=-}^{+} \alpha_{xc}|c\rangle \right)$$

$$= \sum_{x=-\mathcal{N}/2}^{\mathcal{N}/2-1} \alpha_{x+}\hat{T}|x, +\rangle + \alpha_{x-}\hat{T}^\dagger|x, -\rangle$$

$$= \sum_{x=-\mathcal{N}/2}^{\mathcal{N}/2-1} \alpha_{x+}|x + 1, +\rangle + \alpha_{x-}|x - 1, -\rangle, \tag{1.20}$$

where $\hat{T}$ and its conjugate $\hat{T}^\dagger$ are the shift operators (corresponding to $\hat{T}_+$ and $\hat{T}_-$ in Eqs. 1.8 and 1.9), with matrix representation

$$\hat{T} = \begin{pmatrix} 0 & 0 & 0 & 1 \\ 1 & 0 & 0 & 0 \\ 0 & 1 & 0 & 0 \\ 0 & 0 & 1 & \ddots \end{pmatrix}_{\mathcal{N}\times\mathcal{N}} \quad \text{and} \quad \hat{T}^\dagger = \begin{pmatrix} 0 & 1 & 0 & 0 \\ 0 & 0 & 1 & 0 \\ 0 & 0 & 0 & 1 \\ 1 & 0 & 0 & \ddots \end{pmatrix}_{\mathcal{N}\times\mathcal{N}}, \tag{1.21}$$

and $\hat{\mathcal{P}}_+ = |+\rangle\langle+|$ and $\hat{\mathcal{P}}_- = |-\rangle\langle-|$ are projection operators (Townsend 2000a) with matrix representation

$$\hat{\mathcal{P}}_+ = \begin{pmatrix} 1 & 0 \\ 0 & 0 \end{pmatrix} \quad \text{and} \quad \hat{\mathcal{P}}_- = \begin{pmatrix} 0 & 0 \\ 0 & 1 \end{pmatrix}, \tag{1.22}$$

satisfying the identity $\hat{\mathcal{P}}_+ + \hat{\mathcal{P}}_- = \hat{I}$. Hence

$$\hat{T} = \begin{pmatrix} 0 & \hat{\mathcal{P}}_- & 0 & \hat{\mathcal{P}}_+ \\ \hat{\mathcal{P}}_+ & 0 & \hat{\mathcal{P}}_- & 0 \\ 0 & \hat{\mathcal{P}}_+ & 0 & \hat{\mathcal{P}}_- \\ \hat{\mathcal{P}}_- & 0 & \hat{\mathcal{P}}_+ & \ddots \end{pmatrix} \tag{1.23}$$

is the conditional translation operator's $\mathcal{N} \times \mathcal{N}$ block matrix representation. Alternatively the conditional translation operator can, similar to the coin operator, be expressed via a simple mapping given by

$$|x, \pm\rangle \xrightarrow{\hat{T}} |x \pm 1, \pm\rangle.$$

Note that preserving the unitarity of operator $\hat{T}$ requires that it conserves probability, i.e. $\hat{T}|\psi\rangle$ causes the amplitude at position $\mathcal{N}$ in the state vector to wrap around. Nonetheless most physical implementations assume that $\mathcal{N}$ is adequately large and the walk begins sufficiently away from the boundaries of the finite or semi-infinite line that the amplitude in the neighborhood of the boundaries remains negligible throughout the walk duration. Under these conditions the boundary issues arising from the unitarity of the translation operator can be neglected for all practical purposes.

Finally, the full evolution of the quantum walk is given by

$$|\psi_n\rangle = \left(\hat{T}\hat{C}\right)^n |\psi_0\rangle, \tag{1.24}$$

where $|\psi_0\rangle$ and $|\psi_n\rangle$ are the initial and final states of the walk after $n$ steps, and the $2\mathcal{N} \times 2\mathcal{N}$ matrix operator $\hat{U} = \hat{T}\hat{C}$ denotes the evolution operator for a single step of the walk.

It is also possible, and often useful, to perform the quantum walk in the Fourier space, noting that

$$|x, \pm\rangle = \frac{1}{\mathcal{N}} \sum_{k=0}^{\mathcal{N}-1} e^{-i\omega_k x} |k, \pm\rangle, \tag{1.25}$$

where $x = 0, 1, 2 \ldots \mathcal{N} - 1$ and $\omega_k = 2\pi k / \mathcal{N}$. Utilizing the Fourier shift theorem, the translation operator can be reduced to an eigenvalue problem

$$\hat{T}|k, +\rangle = e^{-i\omega_k}|k, +\rangle, \quad \hat{T}^\dagger|k, -\rangle = e^{i\omega_k}|k, -\rangle. \tag{1.26}$$

Taking advantage of the position independent coin operator in Eq. 1.18, we find

$$\hat{U}\mathcal{F}\{|\psi\rangle\} = \hat{U}\sum_{k=0}^{\mathcal{N}-1}\sum_{c=-}^{+}\beta_{kc}|k, c\rangle$$

$$= \sum_{k=0}^{\mathcal{N}-1}\left(|k\rangle \otimes \left(e^{-i\omega_k}\hat{P}_+ + e^{i\omega_k}\hat{P}_-\right)\hat{C}\sum_{c=-}^{+}\beta_{kc}|c\rangle\right)$$

$$= \sum_{k=0}^{\mathcal{N}-1}\left(|k\rangle \otimes e^{-i\sigma_z\omega_k}\hat{C}\sum_{c=-}^{+}\beta_{kc}|c\rangle\right)$$

$$= \sum_{k=0}^{\mathcal{N}-1}\left(|k\rangle \otimes \hat{C}_k\sum_{c=-}^{+}\beta_{kc}|c\rangle\right), \tag{1.27}$$

where $\sigma_z$ is the Pauli $z$ matrix and $\hat{C}_k$ has absorbed the contribution from the projection operators. In other words, the evolution of the quantum walk in Fourier space requires only the application of a modified coin operator $\hat{C}_k$ at every node $k$. Nayak and Vishwan (2000) used this method to give an exact solution for the Hadamard walk on the line. Some results on the symmetry of the Hadamard walk probability distributions are also provided in Konno et al. (2004).

It is important here to note that the conditional shift operator $\hat{T}$ is responsible for generating entanglement between the coin and position degrees of freedom (Abal et al. 2006c). Hence although the walk can be initialized such that the coin and position states are separable, after some steps the coin and position states of the walk become entangled, in general, and are no longer separable. Starting with the position state $|\psi_x\rangle = \cdots a_{-1}|-1\rangle + a_0|0\rangle + a_1|1\rangle \cdots$ and the coin state $|\psi_c\rangle = b_+|+\rangle + b_-|-\rangle$, the product state is given by $|\psi\rangle \equiv |\psi_x\rangle \otimes |\psi_c\rangle = \cdots a_{-1}b_+|-1, +\rangle + a_{-1}b_-|-1, -\rangle + a_0b_+|0, +\rangle + a_0b_-|0, -\rangle + a_1b_+|1, +\rangle + a_1b_-|1, -\rangle \cdots$, where the coefficients $a_x b_c$ are represented by complex values $\alpha_{xc}$ of the vector in Eq. 1.14. Taking the walk in Fig. 1.6 as an example, the state of the walker after one step (Fig. 1.6c) is given by $|\psi\rangle = \mathcal{A}|-1, -\rangle + \mathcal{B}|1, +\rangle$, meaning $a_{-1}b_+ = 0$ while $a_{-1}b_- = \mathcal{A} \neq 0$ and $a_1b_+ = \mathcal{B} \neq 0$ which are mutually exclusive. Hence the state of the walker is no longer separable and cannot be represented as a product state. A number of studies have examined the various aspects of entanglement in quantum walks. Most notably Carneiro et al. (2005) used numerical simulations to study the variation in the entanglement between the coin and position of the particle on regular finite graphs such as cycles. Venegas-Andraca et al. (2005) introduced a variation of the walk on a line using a four dimensional coin obtained from the

entanglement of two qubits. The translation operator was then modified to shift
the walker to the left and right only for states $|00\rangle$ and $|11\rangle$. Abal et al. (2006a)
attempted to quantify the coin-position entanglement generated by the evolution
operator of a discrete-time quantum walk, using the von Neumann entropy of the
reduced density operator (entropy of entanglement). They showed that the entropy
of entanglement converges, in the long time limit, to a well defined value which
depends on the initial state. They also investigated the case of quantum walk on
a plane using two non-separable coins. Finally, Abal et al. (2006c) showed that
for a Hadamard walk with local initial conditions, the asymptotic entanglement
is identical for all initial coin states, while when nonlocal initial conditions are
considered, the asymptotic entanglement varies smoothly between almost complete
entanglement and no entanglement.

### 1.1.4   Walking on Graphs

Quantum walks are often extended beyond their simple 1D predecessor to more
complex graphs with differing degrees of complexity and generalization. A basic
but common extension, for example, is the walk on a circle or cycle, which involves
linking the first and last nodes in the position space and allowing the evolving wave-
function to wrap around. Another straight forward extension of the quantum walk to
a $d$-dimensional lattice (Bach et al. 2004) involves $d$ independent 1D walks prop-
agated by a unitary operator $\hat{\mathcal{U}} = \prod_{i=1}^{d} \hat{T}^{(i)}\hat{C}^{(i)}$, where $\hat{C}^{(i)}$ and $\hat{T}^{(i)}$ are the usual
coin and translation operators propagating the walk on a line in the $i$th dimension.

There has been a considerable body of work exploring the dynamics of quantum
walks on various graphs. Most notably, the $\mathcal{N}$ cycle has been treated in Aharonov
et al. (2001) and Tregenna et al. (2003), and the hypercube in Shenvi et al. (2003),
Moore and Russell (2002), Kempe (2003a), Krovi and Brun (2006a) and Krovi and
Brun (2006b), while Krovi and Brun (2007) provided a treatment of quotient graphs.
Quantum walks on general undirected graphs are defined in Kendon (2006a),
Watrous (2001) and Ambainis (2003), and on directed graphs in Montanaro (2007).

Formally a graph can be defined as a set of $\mathcal{N}$ vertices $v$ and a set of edges $e$
joining pairs of vertices. An edge joining the vertices $v_x$ and $v_y$ is *undirected* if
$e_{xy} \equiv e_{yx}$. The degree $d_x$ of a vertex $v_x$ is the number of edges that are connected
to it. Assuming that there is at most one edge between any pair of vertices, each
vertex can have a maximum degree of $\mathcal{N}$, including a self loop connecting the vertex
back to itself. Hence the maximum number of edges the graph could have in total is
$\mathcal{N}(\mathcal{N}+1)/2$, one between every possible pair of vertices including a single self loop
per vertex. A $d$-regular graph is one in which all vertices have the same degree $d$.
All graphs can be conveniently expressed using an $\mathcal{N}\times\mathcal{N}$ adjacency matrix (Kendon
2006a; Kempe 2003b) with elements

$$A_{xy} = \begin{cases} 1 \text{ if vertices } v_x \text{ and } v_y \text{ are connected by an edge} \\ 0 \text{ otherwise.} \end{cases} \qquad (1.28)$$

**Fig. 1.11** A graph with
various degrees and a labeling
of the edges for each vertex.
Adding self-loops enables the
formation of a four-regular
graph

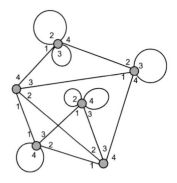

While in general there is no unique way of defining an extension to the quantum
walk on a line, several formulations of the coined quantum walk on a general graph
have been proposed (Ambainis 2003). In most approaches the vertices of the graph
form the position states of the walk labeled $|v_x\rangle$ or more conveniently $|x\rangle$ where $x =
1 \ldots \mathcal{N}$, and each position state has $d_x$ internal or coin states $|c\rangle$ for $c = 1 \ldots d_x$.
This resembles a walker at an intersection $x$ with $d_x$ connecting pathways and a
$d_x$-sided coin which the walker flips to decide which path to take. As before, the
defining feature of the quantum walk is the preservation of the walker's position and
coin in a superposition of states throughout the walk. Hence analogous to Eq. 1.13
we can define the complete state of the system as

$$|\psi\rangle = \sum_{x=1}^{\mathcal{N}} \sum_{c=1}^{d_x} \alpha_{xc} |x, c\rangle, \tag{1.29}$$

where $\alpha_{xc} = \langle x, c | \psi \rangle$ is the amplitude to be at vertex $|x\rangle$ and in the coin state $|c\rangle$.
Similar to the quantum walk on a line, the global coin operator $\hat{C}$ is represented by a
block diagonal matrix (Eq. 1.15), where each block is in turn a $d_x \times d_x$ unitary matrix
representing a local coin operator $\hat{C}_x$. Extending Eq. 1.17, the action of the local coin
operator on a walker in position state $|x\rangle$ and coin state $|i\rangle$ can be represented by
the mapping

$$|x, i\rangle \xrightarrow{\hat{C}_x} \sum_{j=1}^{d_x} \gamma_{ji} |x, j\rangle, \tag{1.30}$$

where $\gamma_{ji}$ are the entries in the $i$th column of $\hat{C}_x$. Clearly if the graph is not regular,
local coin operators have different dimensions for vertices with different degrees.
A work around is possible however by constructing a corresponding regular graph,
where one or more self-loops are added to each vertex of degree less than $d_{\max}$ as
depicted in Fig. 1.11 (Kempe 2003b). All local coin operators associated with the
resulting graph will then have $d_{\max} \times d_{\max}$ dimensions.

Let us now consider an edge $e_{xy}$ on the graph connecting the vertices $x$ and $y$. In
traversing this edge we are at liberty to define the mapping due to the translation
operator $\hat{T}$ from a range of possibilities for connecting one of the initial states

$|x, 1\rangle \ldots |x, d_x\rangle$ to one of the final states $|y, 1\rangle \ldots |y, d_y\rangle$. To systematize this process, Watrous (2001) suggested labeling the coin states of each vertex using the edges connected to it. In other words

$$|\psi\rangle = \sum_{x=1}^{\mathcal{N}} \sum_y \alpha_{xy} |x, e_{xy}\rangle, \tag{1.31}$$

where $y$ represents $d_x$ other vertices $x$ is connected to. Watrous (2001) then defined $\hat{T}$ as a conditional *swap* operator where

$$\hat{T}|x, e_{xy}\rangle = |y, e_{yx}\rangle, \tag{1.32}$$

or equivalently

$$\hat{T} = \sum_{x,y} |y, e_{yx}\rangle \langle e_{xy}, x|, \tag{1.33}$$

for all connected pairs of vertices $x$ and $y$. While Watrous's proposal (2001) can be directly implemented on any irregular graph, a similar approach by Kendon (2006a) involves constructing an equivalent $\mathcal{N}$-regular graph by redefining the action of the conditional swap operator $\hat{T}$, such that

$$\hat{T}|x, y\rangle = \begin{cases} |y, x\rangle \text{ if vertices } x \text{ and } y \text{ are connected (i.e. } A_{xy} = 1) \\ |x, y\rangle \text{ otherwise (i.e. } A_{xy} = 0), \end{cases} \tag{1.34}$$

and $\hat{T}|x, x\rangle = |x, x\rangle$ can be identified as a self loop (Fig. 1.12). If necessary, the coin matrix for each vertex can then be designed in such a way as to effectively deactivate the unwanted self-loops at that vertex (for more details see Fig. 3.76). Both these proposals have the advantage of being independent of specific connectivity criteria, and enable the systematic implementation of quantum walks on any arbitrary graph. The walk can then proceed, as before, according to

$$|\psi_n\rangle = (\hat{T}\hat{C})^n |\psi_0\rangle, \tag{1.35}$$

where $\hat{\mathcal{U}} = \hat{T}\hat{C}$ is now an $M \times M$ matrix operator, with $M = \sum_{x=1}^{\mathcal{N}} d_x$. Other generalizations of quantum walks on graphs include the *edge walk* considered by Hillery et al. (2003) (Sect. 3.1.6 for more details).

### 1.1.5 Multi-particle Quantum Walks

In the classical world, a multi-particle random walk is simply a collection of several particles walking independently in random, assuming no interaction between these particles. In the quantum world, however, a multi-particle walk has inherent

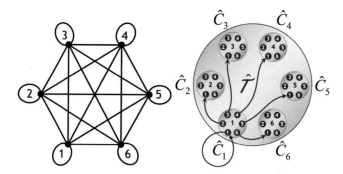

**Fig. 1.12** A 6-regular graph and its corresponding discrete-time quantum walk Hilbert space. Each node or position state $x \in [1, 6]$ consists of 6 internal coin states which undergo unitary transformations induced by the $6 \times 6$ local coin operator $\hat{C}_x$. The conditional translation operator swaps the states $|x, y\rangle$ and $|y, x\rangle$

correlations arising from entanglement and quantum statistics, even if these particles do not physically interact.

The theoretical study of discrete-time quantum walks with more than one particle was initiated by Omar et al. (2006), who considered non-interacting two-particle quantum walks on the infinite line. Omar et al. (2006) established a role for entanglement in two-particle quantum walks by showing that initial states which are entangled in their coin degrees of freedom can generate two-particle probability distributions in which the positions of the two particles exhibit quantum correlations. Likewise, Stefanak et al. (2006) considered the meeting problem in the discrete-time quantum walk of non-interacting particles.

In addition to these studies concerned with distinguishable particles, there have also been theoretical investigations of two-photon quantum walks (Pathak and Agarwal 2007; Rohde et al. 2011). By increasing the dimension of the coin Hilbert space to incorporate both the direction of photon propagation and polarisation, Pathak and Agarwal (2007) introduced a quantum walk in which two photons, initially in separable Fock states, become entangled through the action of linear optical elements. While the quantum walk studied in Omar et al. (2006) and Stefanak et al. (2006) requires entangled initial states to generate spatial correlations, the two-photon walk studied in Pathak and Agarwal (2007), with its larger coin space, is capable of generating entanglement even from initially separable states. Venegas-Andraca (2009) have also proposed a variant of the two-particle quantum walk in which the particles have a shared coin space. In this system, entanglement is introduced between the particles' spatial degrees of freedom by performing measurements in the shared coin space. More recently, Berry and Wang (2011) introduced two spatial interaction schemes, which have shown to dynamically generate complex entanglement between the two walking particles.

As described in Omar et al. (2006) and Berry and Wang (2011), a two-particle quantum walk takes place in the Hilbert space $\mathcal{H} = \mathcal{H}_1 \otimes \mathcal{H}_2$, where $\mathcal{H}_i = (\mathcal{H}_v \otimes \mathcal{H}_c)_i$ for particle $i$, with $v$ and $c$ representing the position and coin space respectively. Denoting the two-particle basis states as

$$|v_i, v_k; c_j, c_l\rangle = |v_i, c_j\rangle_1 \otimes |v_k, c_l\rangle_2, \tag{1.36}$$

we can write the wave-function describing the two-particle system as a linear combination of basis states

$$|\psi\rangle = \sum_{ik}\sum_{jl} a_{v_i,v_k;c_j,c_l} |v_i, v_k; c_j, c_l\rangle. \tag{1.37}$$

The normalization condition is expressible as:

$$\sum_{ik}\sum_{jl} |a_{v_i,v_k;c_j,c_l}|^2 = 1. \tag{1.38}$$

The time evolution operator for the discrete-time two-particle quantum walk is

$$\hat{\mathcal{U}} = \hat{T} \cdot (\hat{I} \otimes \hat{C}), \tag{1.39}$$

where the translation operator $\hat{T} = \hat{T}_1 \otimes \hat{T}_2$ acts on the basis state as

$$\hat{T}|v_i, v_k; c_j, c_l\rangle = |v_j, v_l; c_i, c_k\rangle, \tag{1.40}$$

and the coin operator $\hat{C}_{ik}$ is a $d_i d_k \times d_i d_k$ unitary matrix given by $\hat{C}_{ik} = (\hat{C}_i)_1 \otimes (\hat{C}_k)_2$. Here $d_i$ is the degree of vertex $i$, $(\hat{C}_i)_1$ is the $d_i \times d_i$ coin matrix for the vertex $v_i$, and $(\hat{C}_k)_2$ is the $d_k \times d_k$ coin matrix for the vertex $v_k$. The matrix $\hat{C}_{ik}$ acts on the $d_i d_k$-dimensional coin space with the basis states ordered in the following manner:

$$\{|v_i, v_k; c_1, c_1\rangle, \ldots, |v_i, v_k; c_1, c_{d_k}\rangle, |v_i, v_k; c_2, c_1\rangle, \ldots, |v_i, v_k; c_{d_i}, c_{d_k}\rangle\}.$$

The joint probability of particle 1 located at vertex $v_i$ and particle 2 located at vertex $v_k$ simultaneously, after $n$ steps of the quantum walk is

$$P(i, k, n) = \sum_{jl} |\langle v_i, v_k; c_j, c_l|\hat{\mathcal{U}}^n|\psi_0\rangle|^2. \tag{1.41}$$

The marginal probabilities for particles 1 and 2 are obtained by summing the joint probability over the position states of the other particle, i.e.

$$P_1(i, n) = \sum_k P(i, k, n) = \sum_k \sum_{jl} |\langle v_i, v_k; c_j, c_l|\hat{\mathcal{U}}^n|\psi_0\rangle|^2, \tag{1.42}$$

$$P_2(k, n) = \sum_i P(i, k, n) = \sum_i \sum_{jl} |\langle v_i, v_k; c_j, c_l|\hat{\mathcal{U}}^n|\psi_0\rangle|^2. \tag{1.43}$$

## 1.2  Continuous-Time Quantum Walks

Continuous-time quantum walks were initially proposed by Farhi and Gutmann (1998), out of a study of computational problems reformulated in terms of decision trees. Figure 1.2 is an example of a simple decision tree, where each node of the tree at a given level is connected to two other nodes at the next level and Alice makes a transition to either of these nodes (by stepping to the left or right) with probability 0.5. Classical Markov chains are often employed to study classical random walks through this and other more generalized decision trees (i.e. trees with an arbitrary number of branches fanning out of each node). Suppose we are given a decision tree that has $\mathcal{N}$ nodes indexed by integers $i = 1, \ldots, \mathcal{N}$. We can then define an $\mathcal{N} \times \mathcal{N}$ transition matrix $M$ with elements

$$m_{ij} = \begin{cases} p_{ij} \text{ for } i \neq j \text{ if nodes } i \text{ is connected to node } j \\ 0 \quad \text{otherwise}, \end{cases} \tag{1.44}$$

where $p_{ij}$ is the probability for going from node $i$ to node $j$ satisfying the condition

$$\sum_{i=1}^{\mathcal{N}} p_{ij} = 1 \tag{1.45}$$

for all $j$. Defining $\mathbf{P}$ as the probability distribution vector for the nodes, each step of the random walk through the tree is then given by

$$\mathbf{P}_{t+1} = M \mathbf{P}_t, \tag{1.46}$$

where $t$ represents the integer time step. This is known as a discrete-time Markov process or Markov chain where, the state of the system at $t + 1$ is only dependent on the present state at $t$ and is independent of any past states.

To make the process continuous in time we can assume that transitions can occur at all times and the jumping rate from a node to its neighboring nodes is given by $\gamma$, a fixed, time-independent constant. We may then define the transition *rate* matrix $\mathcal{H}$ with elements

$$h_{ij} = \begin{cases} -\gamma_{ij} \text{ for } i \neq j \text{ if nodes } i \text{ is connected to node } j \\ 0 \quad \text{ for } i \neq j \text{ if node } i \text{ is not connected to node } j \\ S_i \quad \text{ for } i = j \end{cases} \tag{1.47}$$

where $\gamma_{ij}$ is the probability per unit time for making a transition from node $i$ to node $j$ and, for $\mathcal{H}$ to be conservative,

$$S_i = \sum_{\substack{j=1 \\ j \neq i}}^{\mathcal{N}} \gamma_{ij}. \tag{1.48}$$

Notably, in the special case where all transitions occur at the uniform rate $\gamma$, the transition rate matrix is reduced to

$$\mathcal{H} = -\gamma L, \qquad (1.49)$$

where $L = A - D$ is the Laplacian of the graph, $A$ is the graph's adjacency matrix and $D$ is a diagonal matrix whose elements $D_j$ represent the degree of vertex $j$ (Childs and Goldstone 2004b).

Analogous to Eq. 1.46, for any transition rate matrix $\mathcal{H}$, the transitions can be described by a differential equation

$$\frac{d\mathbf{P}(t)}{dt} = -\mathcal{H}\mathbf{P}(t), \qquad (1.50)$$

for which the solution is

$$\mathbf{P}(t) = \exp(-\mathcal{H}t)\mathbf{P}(0), \qquad (1.51)$$

known as the master equation.

Farhi and Gutmann (1998) proposed using the classically constructed transition rate matrix $\mathcal{H}$ to evolve the continuous-time state transitions *quantum mechanically*. This would simply involve replacing the real valued probability distribution vector $\mathbf{P}(t)$ with a complex valued wave vector $\Psi(t)$ and adding the complex notation $\mathtt{i}$ to the evolution exponent, i.e.

$$\Psi(t) = \exp(-\mathtt{i}\hat{\mathcal{H}}t)\Psi(0). \qquad (1.52)$$

The probability distribution vector $\mathbf{P}(t) = |\Psi(t)|^2$, where $\Psi(t)$ is a complex valued vector with elements $\psi(i,t)$ representing the quantum walker's amplitude for landing on node $i$ of the decision tree at time $t$.

It is evident from the above formulation that, unlike the discrete-time quantum walk, quantum states of the continuous-time walk do not incorporate any coin states and hence there are no coin operators required for evolving the walk. Instead the continuous-time walk relies entirely on the node-to-node transition of complex amplitudes. Moreover, given the one-to-one correspondence between the walk's transition (rate) matrix and the adjacency matrix of an arbitrary graph, the above formulation of the quantum walk is not limited to decision trees and can be trivially applied to any general undirected graph with $\mathcal{N}$ vertices. As an example Xu (2008) considered the continuous-time quantum walk on a 1D ring lattice of $\mathcal{N}$ nodes in which every node is connected to its $2m$ nearest neighbors ($m$ on either side).

Similar to the discrete-time quantum walk however, the continuous-time quantum walker can simultaneously exist at all possible position states or nodes of the tree with an amplitude $\psi(i,t)$ as long as its position state remains unmeasured. Here too, it is the interference of these complex amplitudes, as the walk evolves, which give the quantum walk its non-classical propagation characteristics. The two plots in Fig. 1.13 show the characteristic probability distribution for continuous-time quantum walks on a line, with $\mathcal{N} = 150$ nodes indexed as $i = -75 \ldots 75$. In both cases the system starts with an initial state $\psi(i = 0, t = 0) = 1$ and we have plotted

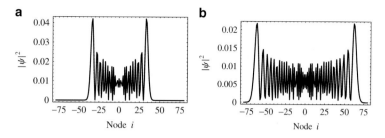

**Fig. 1.13** The probability distributions of two continuous-time quantum walks on a line at $t = 18$ and transition rate matrices given by Eqs. 1.53 and 1.54 respectively. The wave-function spreads more rapidly as additional connections with non-neighboring nodes are included in the transition rate matrix

the probability distribution for $|\Psi(t = 18)|^2$. The first plot represents a walk on a line, where each node is connected to its neighboring nodes by a constant transition rate $\gamma = 1$, resulting in a transition rate matrix given by

$$\hat{\mathcal{H}} = \begin{pmatrix} -2 & 1 & 0 & 0 \\ 1 & -2 & 1 & 0 \\ 0 & 1 & -2 & 1 & \cdots \\ 0 & 0 & 1 & -2 \\ & & & \vdots & & \ddots \end{pmatrix}. \tag{1.53}$$

The second plot on the other hand shows the probability distribution associated with a quantum walk in which the walker makes transitions from each node to five neighboring nodes on each side, making

$$\hat{\mathcal{H}} = \begin{pmatrix} \ddots & & \vdots & & & & & & \\ & \gamma_{-5} & \cdots & \gamma_0 & \cdots & \gamma_5 & 0 & 0 & 0 \\ \cdots & 0 & \gamma_{-5} & \cdots & \gamma_0 & \cdots & \gamma_5 & 0 & 0 \\ & 0 & 0 & \gamma_{-5} & \cdots & \gamma_0 & \cdots & \gamma_5 & 0 & \cdots \\ & 0 & 0 & 0 & \gamma_{-5} & \cdots & \gamma_0 & \cdots & \gamma_5 \\ & & & & & \vdots & & & \ddots \end{pmatrix}, \tag{1.54}$$

where the transition rates $\gamma_{-5} = \gamma_5 = 3.17 \times 10^{-4}$, $\gamma_{-4} = \gamma_4 = 4.96 \times 10^{-3}$, $\gamma_{-3} = \gamma_3 = 3.96 \times 10^{-2}$, $\gamma_{-2} = \gamma_2 = 0.238$ and $\gamma_{-1} = \gamma_1 = 1.66$ are chosen to be the coefficients of the finite difference approximation to the second derivative.

A comparison between the probability distributions of discrete- and continuous-time quantum walks on a line (Figs. 1.7 and 1.13), reveals similar and characteristic propagation behavior (Kempe 2003b; Patel et al. 2005). Moreover given the well established connection between the discrete- and continuous-time classical random walks presented above (i.e. in the classical world, continuous-time walk can be obtained as a limit of the discrete-time walk), one would naturally expect a similar connection between their quantum counterparts. This however turns out to be a nontrivial problem. Childs' work (2010) is among the latest in a line of theoretical efforts to establish a formal connection between the two classes of quantum walks (Strauch 2006a; Konno 2005; Patel et al. 2005).

Manouchehri and Wang (2007) investigated the role of orthogonal position states in the evolution of a continuous-time quantum walk, demonstrating that orthogonality of position states is indeed a subtle but important requirement for the characteristic quantum walk propagation. In Farhi and Gutmann's (1998) treatment of the quantum walk, an arbitrary graph with $\mathcal{N}$ vertices can be represented as $\mathcal{N}$ *position states* with coordinate vectors $|\vec{x}_i\rangle$, for $i = 1, 2 \ldots \mathcal{N}$. These state vectors form an orthonormal basis in the $\mathcal{N}$-dimensional Hilbert space $\mathcal{S}$, that is $\langle \vec{x}_i | \vec{x}_j \rangle = \delta_{ij}$ and the wave-function remains normalized. The time evolution of the quantum walk can be considered as continuously displacing the walker (in time) by a distance $\ell_{ij}$ from node $i$ to all its neighboring nodes $j$ at the rate $\gamma_{ij}$, where $\ell_{ij} = \left\| \vec{x}_i - \vec{x}_j \right\|$ is defined as the transition length. Since the quantum walker can only be present at positions $\vec{x}_1, \vec{x}_2 \ldots \vec{x}_{\mathcal{N}}$, the state space is said to be discrete and the walker has an infinitely narrow width. In other words, as illustrated in Fig. 1.14a, there is no uncertainty or distribution associated with the amplitude $\psi(i, t) = \langle \vec{x}_i | \psi(t) \rangle$ represented as narrow lines over each node. Clearly in this discrete model the nodes can be made arbitrarily close by making $\ell \longrightarrow 0$ without affecting the outcome in any way. The situation changes however when the state space is continuous, meaning that vector $\vec{x}$ is no longer restricted to coordinates $\vec{x}_1, \vec{x}_2 \ldots \vec{x}_N$ and the Hilbert space is an infinite dimensional continuum $\mathcal{S}'$. A consequence of this is that the quantum walker's position at $\vec{x}_i$ can now have a finite uncertainty associated with it. This may be conveniently represented as a Gaussian distribution with a finite width $\Delta x$ that is centered at $\vec{x}_i$ as depicted in Fig. 1.14b, where the amplitude $\psi(i, t)$ is given by the area under the distribution.

Manouchehri and Wang (2007) highlighted the relationship $\ell \gg \Delta x$ as a necessary condition for the quantum walk to retain its characteristic features, meaning that quantum walks require a discrete or orthonormal state space. More specifically Manouchehri and Wang (2007) considered a continuous-time continuous-space quantum walk on a line, propagated using an arbitrary transition rate matrix $\mathcal{H}$ defined as the finite element representation of the operator $-\frac{1}{2}\nabla^2$. They showed that when a finite width walker conforms to the relationship $\ell \gg \Delta x$, the resulting propagation displays the characteristic quantum walk signature as expected. But when the transition length approaches a continuum, i.e. $\ell \longrightarrow 0$ while keeping $\Delta x$ unchanged (Fig. 1.14c), the propagation behavior begins to alter until it converges to that of a conventional quantum wave in free space. Figure 1.15 shows this transition for a quantum walk on a quasi-continuous line approximated by a numerical vector.

Beyond theoretical interest however, this analysis finds important implications for the physical implementation of continuous-time quantum walks. Indeed a number of existing experimental schemes implicitly verify the notion that state

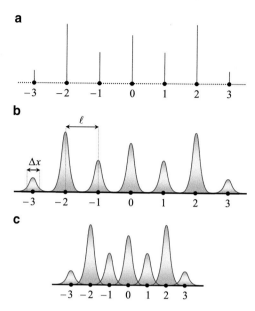

**Fig. 1.14** (**a**) Continuous-time quantum walk on a line of discrete nodes. *Vertical lines* represent the walker's probability amplitude to be at each node. (**b**) Continuous-time quantum walk on a continuous line. The nodes of the walk are a distance $\ell$ apart and localized distributions represent the walker's probability amplitude to be at each node with an associated uncertainty $\Delta x$. The condition $\ell \gg \Delta x$ means that the nodes are well separated and their respective distributions do not interfere. (**c**) The nodes of the walk have been moved closer such that $\ell \sim \Delta x$. Consequently the localized distributions at neighboring nodes begin to overlap and interfere with one another

space should be discrete. Du et al. (2003) demonstrated the implementation of a continuous-time quantum walk on a circle with four nodes, using a two-qubit NMR quantum computer. Here the four dimensional state space spanned by the spin states of the two qubits $|\uparrow\uparrow\rangle, |\uparrow\downarrow\rangle, |\downarrow\uparrow\rangle$ and $|\downarrow\downarrow\rangle$, is clearly discrete in nature and the walker has an infinitely narrow width. Nonetheless the assumption that the walker has a finite width is in fact a realistic one for many other physical systems such as a single particle in real or momentum space, where a localized distribution invariably arises from the fundamental uncertainty in the particle's position for any given coordinates. In one such proposal Côté et al. (2006) described a scheme based on ultra cold Rydberg (highly excited) $^{87}$Rb atoms in an optical lattice. In this scheme the walk is taking place in real space which is clearly continuous, but the confinement of atomic wave-function to individual lattice sites amounts to a virtually-discrete state space. In fact Côté et al. (2006) prescribed an even more conservative condition: to eliminate the atoms except in every fifth site (spacing $25\,\mu$m) in order to achieve a better fractional definition of the atom separation. Similarly Solenov and Fedichkin (2006a) proposed using an a ring shaped array of identical tunnel-coupled quantum dots to implement the continuous-time quantum walk on a circle. Given the confinement of the electron wave-function inside

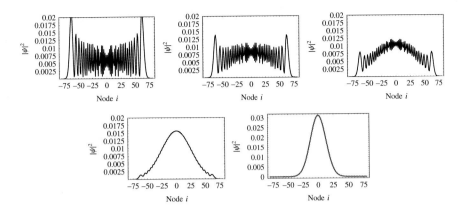

**Fig. 1.15** Transition of the continuous-time quantum walk probability distribution from its characteristic signature to a simple Gaussian for discrete transition lengths $\ell = 16, 4, 3, 2$, and 1. The transition rate matrix $\mathcal{H} \approx -\frac{1}{2}\nabla^2$ (given by the 10th order finite difference approximation) propagates the walk on a line of $\mathcal{N} = 160$ nodes labeled $i = 0, \pm 1, \pm 2, \ldots$. The simulation is carried out using a numerical vector with $\mathcal{N} \times \ell$ elements labeled $x$ where each node represents a block of $w = 16$ elements centered at equidistant locations $x_i$ such that $x_i - x_{i-1} = \ell$. The state of the walker at time $t$ is given by $\psi(i,t) = \int_{-w/2}^{w/2} \mathcal{G}_i(x,t)\, dx$ where $\mathcal{G}_i(x,t) = \mathcal{A}_i(t) \exp\left((x - x_i)^2/(2\Delta x)^2\right)$ is a Gaussian distribution with $\Delta x = 2$ and a complex phase $\mathcal{A}_i(t)$. Each plot shows the final probability distribution $|\psi(i, t = 15)|^2$ arising from an initial Gaussian wave-packet at the center node with $|\psi(i = 0, t = 0)|^2 = 1$. The *dotted line* on the last plot represents the analytical solution for the quantum propagation of the initial wave-packet in free space over the same time period

the individual quantum dots, the authors have once again implicitly described a virtually-discrete state space over the continuous real space.

Mülken and Blumen (2011) reviewed recent advances in continuous-time quantum walks and their application to transport in various systems. In particular the work provides an overview of different types of networks on which continuous-time quantum walks have been studied so far. Extensions of continuous-time quantum walks to systems with long-range interactions and with static disorder are also discussed as well as an outlook on possible future directions.

## 1.3    Walking Characteristics

In studies of quantum walks in a finite state space, i.e. a cycle or other more generalized graphs, it is often important to consider a walker with an initially localized probability distribution and then analyze the time it takes for the distribution to spread throughout all nodes. Most authors (Aharonov et al. 2001; Moore and Russell 2002; Kendon and Tregenna 2003; Adamczak et al. 2003) have noted however that unlike the classical random walk, which converges to a uniform distribution in the long time limit, in general a unitary reversible process such as that of evolving the quantum walk does not converge to any stationary distribution at large times. Nevertheless by suitably relaxing the definition, Aharonov et al. (2001) obtained a measure of how fast the quantum walk spreads or how confined the quantum walk stays in a small neighborhood which they quantified as *filling time*, *dispersion time* and most notably *mixing time*.

More specifically, mixing time is used to describe two types of processes. The first one, called *instantaneous mixing*, has arisen from the study of continuous-time quantum walks and refers to the uniform or nearly uniform spread of probability distribution that can happen at some particular moment of evolution (Adamczak et al. 2003; Moore and Russell 2002). The other, *average mixing*, is the decay of the time-averaged deviation of the probability distribution from the uniform case (Kendon and Tregenna 2003).

The time averaging of the probability distribution is given by

$$\overline{P(x,T)} = \frac{1}{T} \sum_{t=0}^{T} P(x,t), \tag{1.55}$$

and is required in order to ensure that coherent probability oscillations converge to a static distribution. Here $T$ is an integer representing the number of steps when considering a discrete-time walk, and a real valued number in the context of a continuous-time walk. The average mixing time is then defined as

$$M_\epsilon = \min\{T \mid \forall\, t > T \;:\; \left\| \overline{P(x,t)} - P_u \right\| < \epsilon\}, \tag{1.56}$$

where $P_u$ is the limiting (uniform) distribution over the cycle or graph.

It can be shown that for an unbiased Hadamard walk on the line, the walk is almost uniformly distributed over the interval $\left[-t/\sqrt{2}, t/\sqrt{2}\right]$ after $t$ time steps (Nayak and Vishwan 2000). This implies that the same walk defined on the circle mixes in linear time, in direct contrast with the quadratic mixing time for the corresponding classical walk. Moore and Russell (2002) analyzed both continuous- and discrete-time quantum walks on the hypercube and showed that in both cases, the walk has an instantaneous mixing time at $n\pi/4$. This has a complexity of $O(n)$ as compared to the classical walk on the hypercube which mixes in time $\Theta(n \log n)$, meaning that the quantum walk is faster by a logarithmic factor. Extending this study, Marquezino and Portugal (2008) showed that the asymptotic distribution of a discrete-time quantum walk on the hypercube is in general not uniform, and characterized the average mixing time to this distribution.

Another quantity of interest in the analysis of random walks is the *hitting time*; the average time for a walk beginning on a particular starting vertex to arrive for the first time at a predetermined target vertex or group of vertices. A discussion of the hitting time is provided in the work of Krovi and Brun (2006a) where they derived an expression for quantum hitting times using superoperators and numerically evaluated it for the discrete-time walk on the hypercube. This is followed by an examination of the dependence of hitting times on the type of unitary coin. While hitting times for a classical random walk on a connected graph will always be

finite, Krovi and Brun (2006b) showed that, by contrast, quantum walks can have
infinite hitting times for some initial states. The authors also established criteria to
determine if a given walk on a graph will have infinite hitting times. In a related
work, Krovi and Brun (2007) described the construction of a quotient graph for
any subgroup of the automorphism group and provided a condition for determining
whether the quotient graph has infinite hitting times given that they exist in the
original graph.

Abal et al. (2006b) defined the survival probability of a discrete-time quantum
walk on a line as

$$P_{surv}(n) = \sum_{j=-s}^{s} P_j(n),$$ (1.57)

where $n$ is the discrete time steps, $s$ is a non-negative integer and the walk is
assumed to proceed in a symmetric range $[-s, s]$. Assuming $P_{surv}(0) = 1$, the
decay of this quantity reflects how fast the walker leaves the region where it is
initially located. For a classical random walk, the corresponding quantity decays
as $n^{-1/2}$. For a quantum walk with an arbitrary coin, initialized at a single site,
the survival probability decays faster than in the classical case, namely as $n^{-1}$.
Abal et al. (2006b) further showed that, if non-local initial conditions with specific
relative phases are considered, e.g. $|\psi_0\rangle = \frac{1}{2}(|L\rangle + i|R\rangle) \otimes (|-k\rangle + i|k\rangle)$, where
$k \leq s$, $P_{surv}(0) = 1$ and only sites $x = |\pm k\rangle$ are initially occupied, then an
enhanced decay rate $n^{-3}$ can be obtained.

## 1.4 Decoherence

In the context of quantum walks, decoherence refers to the transition of the
evolution dynamics from quantum to classical by introducing external "defects" in
the mechanics of the walk. This is commonly quantified by measuring the drift in the
spreading characteristics of the quantum walk distribution, away from the ballistic
(quantum) and towards a diffusive (classical) regime. Several studies on decoherent
quantum walks in one (Kendon and Tregenna 2002; Brun et al. 2003a,b,c; Shapira
et al. 2003; López and Paz 2003; Romanelli et al. 2005) or more dimensions
(Oliveira et al. 2006; Alagić and Russell 2005; Košík et al. 2006) have advanced
our understanding of (a) the relationship between the quantum and classical random
walks; (b) the effects of interaction with the environment particularly in the context
of physical implementations; and (c) ways to control and even enhance the characteristic behavior of the walk. Kendon (2007) provided a substantive review of the
subject for the discrete- as well as the continuous-time quantum walks with a formal
discussion of the non-unitary evolution arising from various forms of decoherence
or measurement. A treatment of quantum to classical decoherence in the context of
the complementarity principle is provided in Kendon and Sanders (2005).

In a key study, Brun et al. (2003a) considered two possible routes to classical behavior for the discrete-time quantum walk on a line by (a) introducing *pure* (unitary) dephasing, whereby for a given parameter $\theta$ there is an equal chance for the coin operator to be dephased either by the operator $\hat{A}_0 = 1/\sqrt{2}\left(e^{i\theta}|+\rangle\langle+| + e^{-i\theta}|-\rangle\langle-|\right)$ or $\hat{A}_1 = 1/\sqrt{2}\left(e^{-i\theta}|+\rangle\langle+| + e^{i\theta}|-\rangle\langle-|\right)$ at every step of the walk; and (b) employing a $D$-dimensional coin with $D > 2$, to dilute the effects of interference between paths. This latter case with $D = 2^M$ is in fact a generalization of the multicoin model where the quantum walk is driven by $M$ arbitrarily different coins (Brun et al. 2003b) and the walker cycles through the available coins such that after $n$ steps each coin is flipped $n/M$ times. The authors concluded that the multicoin walk retains the quantum quadratic growth of the variance except in the limit, where there is a new coin for every step (i.e. $M = n$), while the walk with decoherent coin exhibits classical linear growth of the variance even for weak dephasing. In another study, Brun et al. (2003c) presented a more detailed study of the decoherence in the quantum coin which drives the walk.

Ribeiro et al. (2004) examined the rich variety of evolutions arising from several types of biased quantum coins, arranged in aperiodic sequences. First considering the quasi-periodic Fibonacci sequence $\hat{C}_n = \hat{C}_{n-1}\hat{C}_{n-2}$ to obtain the coin used at the $n$th step of the walk, the authors showed that choosing any pair of initial coins where $\hat{C}_0 \neq \hat{C}_1$ resulted in a sub-ballistic evolution. Furthermore, they found that by switching to an entirely random sequence of $\hat{C}_0/\hat{C}_1$ for each step of the walk and averaging over several complete iterations, the propagation became diffusive.

As we will see in the next chapter, Broome et al. (2010) and Schreiber et al. (2011) have devised novel experimental demonstrations of *controlled* decoherence due to pure (unitary) dephasing of the coin in a one dimensional quantum walk on a line. Broome et al. (2010) also investigated the effect of absorbing boundaries and showed that decoherence significantly affects the probability of absorption as predicted in Ambainis et al. (2001) and Bach et al. (2004). Schreiber et al. (2011) provided another useful formalism for decoherence by considering a quantum walker $|\psi(x)\rangle$ in the position state $|x\rangle$ and coin state $|c_x\rangle$ and expressing its evolution as a recursive sequence given by

$$|\psi(x)\rangle_{n+2} = \gamma|\psi(x)\rangle_n + \beta_{+2}|\psi(x+2)\rangle_n + \beta_{-2}|\psi(x-2)\rangle_n, \qquad (1.58)$$

where $n$ is the step number, and constant matrix coefficients $\gamma$ and $\beta$ are fully set by the local coin operator $\hat{C}$ and can be readily determined by computing $(\hat{T}.\hat{C})^2$ using matrix representations in Eqs. 1.15 and 1.23. In this formalism, components of $\gamma$ and $\beta$ account for the particle's probability of staying at the discrete position $x$ or evolving to an adjacent site $x \pm 2$. In order to introduce decoherence into the system Schreiber et al. (2011) implemented a position dependent coin operator $\hat{C}(x)$ resulting in new position dependent coefficients $\gamma \longrightarrow \gamma_x$ and $\beta_{\pm2} \longrightarrow \beta_{x\pm2}$ that are determined by coin operations $\hat{C}(x)_{n+1}$ and $\hat{C}(x)_{n+2}$. The authors then demonstrated two tunable disorder mechanisms: (a) static disorder introduced by manipulating the coefficients $\gamma_x$ and $\beta_{x\pm2}$ in a manner that is position but not step dependent, leading to the first experimental observation of Anderson localization (Anderson 1958) in a coined quantum walk; (b) dynamic disorder due to randomization of the coefficients to eliminate position correlations, thus simulating the evolution of a particle interacting with a fast fluctuating environment.

While the above studies have explored the effect of altering only the coin degree of freedom, one might naturally consider other modifications of the particle evolution, such as allowing decoherence of the position as well as the coin. There have been numerical studies of this and of the transition to classical behavior that results (Dur et al. 2002; Kendon and Tregenna 2002, 2003).

Abala et al. (2008) considered the effect of different unitary noise mechanisms on the evolution of a quantum walk on a linear chain with a generic coin operation: (a) a coin-flip operation applied with probability $p$ per unit time (i.e. bit-flip channel noise) and (b) at a given time a link on the line is open with probability $p$. This latter (broken-link) model was introduced by Romanelli et al. (2005) as a way to mimic the effects of thermal noise in some experimental situations and was later generalized to higher dimensions by Oliveira et al. (2006). In particular Romanelli et al. (2005) showed that when links between neighboring sites are randomly broken with probability $p$ per unit time, the evolution becomes decoherent after a characteristic time that scales as $1/p$. As noted by the authors, the fact that the quadratic increase of the variance is eventually lost even for very small frequencies of disrupting events suggests that the implementation of a quantum walk on a real physical system may be severely limited by thermal noise and lattice imperfections.

An interesting application of decoherence was demonstrated by Kendon and Tregenna (2003), where they studied the effects of decoherence in the discrete-time quantum walk on a line, cycle, and hypercube. They found that while the walk's sensitivity to decoherence increases with the number of steps, the effect of a small amount of decoherence will in fact enhance the properties of the quantum walk in ways that are desirable for the development of quantum algorithms. More specifically, they observed a highly uniform distribution on the line, a very fast mixing time on the cycle, and more reliable hitting times across the hypercube. Nevertheless Alagić and Russell (2005) proved that quantum hypercube walks possess a decoherence threshold beneath which the essential properties such as linear mixing times are preserved, while beyond this threshold, the walks behave like their classical counterparts. Maloyer and Kendon (2007) showed that the mixing time of a discrete-time quantum walk on an $\mathcal{N}$-cycle may be reduced by allowing for some decoherence from repeated measurements, provided that those measurements affect the position of the walk. Likewise Marquezino and Portugal (2008) established that for a walk on a hypercube a controlled amount of decoherence, due to randomly breaking links between connected sites, helps in obtaining and preserving a uniform distribution in the shortest possible time. Similar effects were also reported by Dur et al. (2002). They introduced two models for studying the experimental uncertainty involved in performing the walk. These two models essentially describe the effect of uncertainty in the quantum coin and the position state of the walker. Richter (2007b) investigated the possibility of gaining an algorithmic speed-up using quantum walks that decohere under repeated randomized measurements (Sect. 2.2). Kendona and Maloyera (2008) examined how the interplay between quantum evolution and random noise or measurements produces optimal computational properties.

While a substantial body of work has investigated decoherence in discrete-time quantum walks, a number of studies have turned their focus on continuous-time quantum walks. Solenov and Fedichkin (2006b) considered the walk on hypercycles and the decoherence arising from continuous week measurements. Keating et al. (2007) applied well-established ideas from the theory of localization (Anderson 1958) to the continuous-time quantum walk of a single particle on a glued-tree and showed that when the graphs have imperfections, as can be the case in a physical implementation, the propagation of quantum information is suppressed exponentially in the amount of imperfection, and is therefore unlikely to be useful for algorithmic purposes. In a related work, Yin et al. (2008) considered the effect of two types of disorder: (a) static disorder which is modeled by a position $x$-dependent random variable; and (b) dynamic for which the random variable is also time $t$-dependent. They concluded that in random media with static disorder, continuous-time quantum walks often perform worse than their classical counterparts due to Anderson localization which effectively stops the propagation of the walk while the corresponding classical walks can still move infinitely slowly. Likewise, they showed that for dynamic disorder there is no benefit from quantum walks, since the ballistic evolution crosses over to classical diffusion after some time and the quantum walk becomes classical.

# Chapter 2
# Potential Applications

As some problems are best solved in classical computation with algorithms based on random walks, it is expected that this type of problems could be solved even faster using quantum walks (Knight et al. 2003b). Kendon (2006b) gave an introductory overview of quantum walk algorithms in the context of prospects for practical quantum computing, particularly in three areas: (a) simulation of classical systems; (b) simulation of quantum systems; and (c) analogue computing. Another review of the subject is provided by Ambainis (2003) who highlighted two main classes of quantum walk algorithms (a) exponentially faster hitting and (b) quantum walk based search. In this chapter we will examine these and a number of other algorithmic applications of quantum walks.

## 2.1  Exponentially Faster Hitting

In this group of algorithms, we have a known graph. We start in one vertex $A$ and would like to reach a certain target vertex $B$ using a random walk process. We take the example of a glued tree (Fig. 2.1) which is constructed by taking two full binary trees of depth $d$ with roots $A$ and $B$ and gluing each leaf of the first tree to a leaf of the second tree. It has $O(2^d)$ vertices and if we start in $A$ and do a classical random walk, the number of steps to reach $B$ will have a lower bound given by $\Omega(d^2)$ (Ambainis 2003). Childs et al. (2003) proposed an algorithm, whereby a continuous-time quantum walk can reach $B$ from $A$ in $O(d^2)$ steps. As noted by Tregenna et al. (2003) however the fast hitting time obtained with a quantum walk is highly sensitive to the symmetry of the problem: for quantum walks starting at a node other than root $A$, the exit becomes exponentially harder to find and the quantum walk does no better than a classical algorithm. An adaptation of the glued tree walk using a discrete-time quantum walk is given by Carneiro et al. (2005) who generalized the glued trees graph to higher branching rate (fan out) and verified that the scaling with branching rate and with tree depth is polynomial (vs. exponential

J. Wang and K. Manouchehri, *Physical Implementation of Quantum Walks*,
Quantum Science and Technology, DOI 10.1007/978-3-642-36014-5_2,
© Springer-Verlag Berlin Heidelberg 2014

**Fig. 2.1** "Glued trees" graph
of Childs et al. (2002)

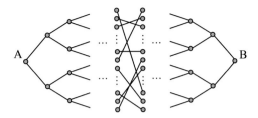

for the classical random walk). Similarly Kempe (2003a) showed that the hitting time of the discrete quantum walk from one corner of a hypercube to the opposite corner is polynomial in the number of steps, $n$, while it is exponential in $n$ in the classical case. Other examples of algorithms of this type include the original work of Farhi and Gutmann (1998) which lead to the introduction of continuous-time quantum walks and another work by Childs et al. (2002).

## 2.2  Quantum Walk Based Search

Search algorithms typically involve an unsorted database containing $\mathcal{N}$ records, of which just one satisfies a particular property. The problem is to identify that one record. Any classical algorithm, deterministic or probabilistic, will take $O(\mathcal{N})$ steps since on the average it will have to examine a large fraction of the $\mathcal{N}$ records. In an early work, Grover (1996) constructed his famous quantum mechanical search algorithm (not using quantum walks) which was capable of finding the record in only $O(\sqrt{\mathcal{N}})$ steps, making use of the fact that quantum mechanical systems can do several operations simultaneously due to their wave like properties.

A similar problem, in the context of quantum walk algorithms, is that of a graph in which one or more vertices are marked and the task is to find the marked vertices. In the case of a coined quantum walk, this is achieved by applying a special "marking coin" to the marked nodes while every other unmarked node is evolved using an unbiased or symmetric coin such as the Grover's coin (Carneiro et al. 2005; Moore and Russell 2002). The query complexity of such a search algorithm is defined as the number of steps required for the quantum walk's initially uniform distribution to converge on the marked node with a high probability. The first search algorithm of this kind was discovered by Shenvi et al. (2003) for searching a Boolean hypercube with $\mathcal{N}$ vertices. While the Grover's algorithm could search this graph in $O(\sqrt{\mathcal{N}} \log \mathcal{N})$ steps (Ambainis 2003), Shenvi et al. (2003) showed how to search it by quantum walk in $O(\sqrt{\mathcal{N}})$ steps.

Childs and Goldstone (2004b) studied quantum search on $d$-dimensional grids, this time using a continuous-time quantum walk. This approach involves modifying the transition rate matrix so that the marked vertex, labeled as $w$, is special. Using $\mathcal{H}_w = -|w\rangle\langle w|$ for example, Childs and Goldstone (2004b) modified the uniform transition rate matrix in Eq. 1.49 such that the resulting rate matrix used for the search is given by $\mathcal{H} = -\gamma L + \mathcal{H}_w$. This yielded an $O(\sqrt{N})$ time quantum algorithm for $d > 4$, an $O(\sqrt{N} \log N)$ time algorithm for the critical dimension $d = 4$ and no speed-up in $d < 4$ dimensions. Investigating the idea that using a quantum coin makes the walk faster, Ambainis et al. (2005) subsequently studied similar search problems using a discrete-time quantum walk. The result was an $O(\sqrt{N})$ time algorithm for $d > 2$ and an $O(\sqrt{N} \log N)$ time algorithm for the critical dimension $d = 2$. Shortly after however, Childs and Goldstone (2004a) matched this improved performance by utilizing a lattice version of the Dirac Hamiltonian which requires the introduction of spin degrees of freedom, thus incorporating additional memory in the continuous-time quantum walk search algorithm. Reitzner et al. (2009) introduced *scattering quantum walks* on highly symmetric graphs and utilized them to solve search problems on these graphs. Using analytical solutions of the walk in modestly dimensioned Hilbert spaces (between three and six dimensions), they found a quadratic quantum speed-up in all cases considered.

Another application of quantum walk search is element distinctness, i.e. given numbers $x_1, \ldots, x_N$, finding $i, j \in [N]$ for $i \neq j$ such that $x_i = x_j$. Classically, element distinctness requires $O(N)$ queries. Ambainis (2004) constructed a quantum walk algorithm which solves the problem with $O(N^{2/3})$ queries.

Richter (2007a,b) proposed a quantum algorithm for *almost uniform sampling* based on the continuous-time quantum walk, analogous to what is used in many classical randomized algorithms. He conjectured that this algorithm could potentially yield quantum speed-up for a number of *fully-polynomial randomized approximation* schemes.

## 2.3 Network Characterization

An interesting phenomenon that arises when studying structures where not all vertices are equivalent is that the successful search probability depends on the location of the marked vertex. Agliari et al. (2010) considered continuous-time quantum walk-based search on fractals, which represents the first effort to characterise the search procedure on structures that are not vertex-transitive. In their study of quantum search on Cayley trees, T-fractals and dual Sierpiński gaskets, Agliari et al. (2010) assumed that a peripheral vertex would be more difficult to find than a more central vertex, i.e. the maximum success probability for a central vertex would be greater than for a peripheral vertex. Soon after, Berry and Wang (2011) analyzed this idea in more detail by studying how the maximum success probability varies with the centrality of the marked vertex. They found that in some simple cases, the maximum success probability does indeed increase with increasing centrality. However they showed that, in general, such a relationship does not hold.

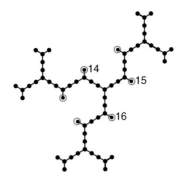

**Fig. 2.2** The third generation regular hyperbranched fractal of functionality, $f = 3$ (RHF$_{3,3}$). The *cul-de-sac* vertices are highlighted (From Berry and Wang 2011)

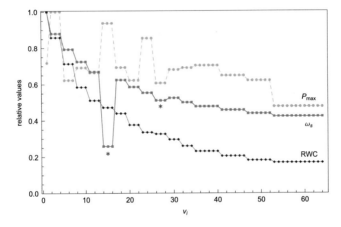

**Fig. 2.3** Numerical results for $P_{\max}$, $\omega_s$ and the classical random walk centrality RWC on RHF$_{3,3}$. All data is normalised for comparison and vertices $v_i$ are ordered with decreasing random walk centrality. The (∗) denotes sets of *cul-de-sac* vertices (From Berry and Wang 2011)

The efficiency of quantum walk-based search relative to classical search is not only determined by the maximum success probability $P_{\max}$, but also the time taken to reach the maximum. Berry and Wang (2011) also analyzed the lowest frequency $\omega_s$ of the success probability as an indicator of the time complexity of the search. Their results suggest that this frequency is correlated with the centrality for a larger class of graphs than the maximum success probability and they discussed exceptions in terms of local structure of these graphs. They demonstrated that maximum success probability and its frequency are determined by the global structure of the graph as well as the centrality of and local structure surrounding the marked vertex, as shown in Figs. 2.2 and 2.3.

## 2.4   Graph Isomorphism

Two graphs $G$ and $H$ are *isomorphic* if there exists a bijection $f : V(G) \rightarrow V(H)$ such that $v_i$ and $v_j$ are adjacent in $G$ if and only if vertices $f(v_i)$ and $f(v_j)$ are adjacent in $H$. Such a mapping $f$ is called an *isomorphism*. In other words, two graphs are isomorphic if the vertices can be relabelled so that the graphs are identical. The question of determining whether two given structures (either algebraic or combinatorial) are isomorphic has been a long-standing open problem in mathematics and computer science. Considerable and continuing effort has been devoted to the graph isomorphism (GI) problem, due to both the variety of practical applications, and its relationship to questions of computational complexity. While efficient GI algorithms exist for certain restricted classes of graphs, there is no known algorithm with polynomial bounds to solve GI for general graphs. The best known general algorithm has an upper bound of $O(e^{\sqrt{N \log N}})$ (Babai et al. 1983; Köbler 2006).

Several papers in the literature (Childs et al. 2003; Kempe 2003b; Gerhardt and Watrous 2003) suggested to study graph isomorphism using quantum walks instead of classical random walks. Although the mathematical formulation for the classical and quantum walks is very similar, they display remarkably different walking characteristics. The classical random walk is a diffusive process and the system always converges to a steady state solution for all graphs. Consequently, introducing some difference in the graph makes little difference in the overall appearance of the probability distribution as shown in Fig. 2.4 (middle panel). On the other hand, quantum walks are a unitary process and completely reversible. They will not diffuse into a steady state, but rather the wave-function amplitude at each node oscillates in such a way that reflects upon the topological structure of the graph. The amplitude distribution is significantly different for even slightly modified graphs, as illustrated in Fig. 2.4 (bottom panel).

An important question then arises: can we identify graph isomorphism from the amplitude distributions resulting from quantum walks on graphs? There have been several attempts to develop quantum walk based GI algorithms in recent years. Shiau et al. (2005) performed single-particle continuous-time quantum walks on closed graphs and concluded that such walks fail to identify non-isomorphic strongly regular graphs (SRGs). A SRG with parameters $(N, d, \lambda, \mu)$ is an undirected graph on $N$ vertices in which each vertex has degree $d$, each pair of adjacent vertices have exactly $\lambda$ common neighbours and each pair of non-adjacent vertices have exactly $\mu$ common neighbours (Godsil and Royle 2001). There is no known polynomial-time GI algorithm for the strongly-regular graphs (SRGs). In the same work the authors also introduced a modified quantum walk algorithm based on two interacting particles which distinguished a set of strongly regular graphs with up to 29 nodes.

One class of classical GI algorithms relies on the eigenvalues of Laplacian matrices, but many non-isomorphic graphs are co-spectral using the Laplacian matrices or some other modified matrix representation (He et al. 2005). Emms et al. (2006)

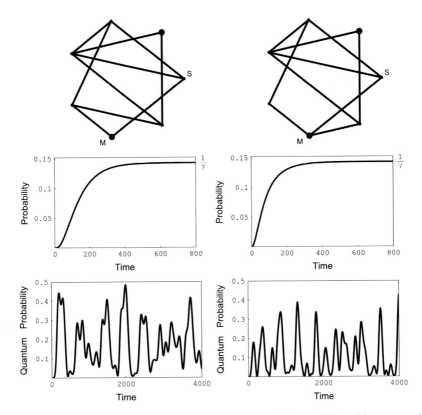

**Fig. 2.4** *Top panel*: randomly generated graph and a slight variation; *middle* and *bottom panels*: probability at node *M* with the walker starting at node *S* for classical and quantum walks, respectively

introduced a new matrix representation inspired by quantum walks. This new matrix displays different spectra for a set of non-isomorphic strongly regular graphs, but fails to distinguish fairly simple graphs as discussed in their paper.

Using discrete-time quantum walks, Douglas and Wang (2008) developed a single particle GI algorithm that successfully distinguished all pairs of graphs they tested, including all strongly regular graphs with up to 64 vertices. In order to generate probability distributions which differ for non-isomorphic SRGs of the same family, found that it was necessary to perturb the quantum walks by applying a phase shift to one vertex of the graph at each time step. Apart from strongly regular graphs, they also tested on trees, planar graphs, projective planes, Eulerian graphs, Hypohamiltonian graphs, vertex critical graphs, edge critical graphs, vertex-transitive graphs, and other regular graphs. All graphs in each category with the same parameters are compared pair-wise and this algorithm has successfully identified all isomorphic and nonisomorphic pairs. Particularly worth mentioning is the set of strongly regular graphs with parameters $(36, 15, 6, 6)$ of which all $529, 669, 878$ pairs were tested and distinguished.

Gamble et al. (2010) examined further the continuous-time two-particle quantum walk on SRGs and found that the dynamics of *non-interacting* two-particle quantum walks is completely determined by the SRG family parameters $(N, d, \lambda, \mu)$. These walks cannot, therefore, be used to distinguish non-isomorphic SRGs from the same family. By introducing an *interaction* between the two particles however, these walks were able to successfully distinguish all non-isomorphic SRGs tested (Gamble et al. 2010).

Quantum-walk based GI algorithms are, in some respects, similar to a classical GI algorithm known as Weisfeiler-Leman (W-L) vertex refinement, which typically distinguishes pairs of non-isomorphic graphs in polynomial time. In fact, recent studies have suggested a direct correspondence between the two methods (Douglas 2011). It has been shown by Cai et al. (1992) however that for certain graphs, there exist two related graphs, which cannot be distinguished by the W-L method in polynomial-time, thus proving the W-L method is not a polynomial-time algorithm for GI.

To test whether or not the interacting two-particle quantum walk procedure can distinguish the Cai-Fürer-Immerman graphs, Berry and Wang (2011) constructed such a pair with 80 vertices and computed their GI certificates. Both graphs in the non-isomorphic pair produced the same certificate, demonstrating that the procedure, in its current form, cannot distinguish arbitrary graphs. Nonetheless, this technique provides an interesting tool for computer scientists interested in GI. In particular, recent work by Douglas (2011) has shown that the Weisfeiler-Leman method can however be extended to distinguish all known counterexamples. It is therefore possible that a quantum-walk based graph isomorphism testing procedure could also be extended in a similar fashion.

## 2.5 Modeling Quantum Phenomena

There is growing interest in the simulations of quantum phenomena as a significant branch of research in quantum computing. Strauch (2006b) investigated the relation between the 1D Dirac equation and quantum walks. He showed through simulations that the time evolution of the probability density of a quantum walker, initially localized on a lattice, is directly analogous to relativistic wave-packet spreading. Later, Bracken et al. (2007) confirmed, via a more detailed analysis, that this is not a coincidence and that the evolution of any positive-energy state of a free Dirac particle moving in one dimension can be modeled arbitrarily closely as a quantum walk on a line, by making the position steps $\Delta t \longrightarrow 0$ and the number of iterations $n \longrightarrow \infty$. This relationship between the Dirac particle evolution and a quantum walk leads, as the authors pointed out, to the intriguing speculation that at some small space-time scale, there may really be a quantum walk defining the evolution of the relativistic electron states and that the Dirac evolution may only be a large-scale approximation.

## 2.6  Universal Computation

In one of the earliest papers on quantum computation, Feynman showed how to implement any quantum circuit by constructing a corresponding time-independent Hamiltonian (Feynman 1985). Recalling the transition rate matrix in Eq. 1.47, it becomes apparent that the dynamics of any time-independent Hamiltonian can be viewed as a continuous-time quantum walk on a *weighted* graph. Childs (2009) presented an alternative Hamiltonian for universal quantum computation where the edges are unweighted, corresponding directly to a continuous-time quantum walk, and the graph has a maximum degree 3. The main idea is to implement quantum gates via scattering processes on graphs. More specifically an $n$-qubit circuit is represented by $2^n$ *virtual* quantum wires and quantum gates are implemented by scattering off widgets attached to and connecting the wires. Childs (2009) then described how to implement such widgets for a universal set of quantum gates, such as the CNOT gate and two single-qubit gates that generate SU(2), by scattering on graphs.

This result is one of the strongest indications so far that the continuous-time quantum walk is computationally powerful: in principle, any quantum algorithm can be recast as a quantum walk algorithm. More precisely, any $m$-gate quantum circuit can be simulated by a simple quantum walk on an $N$-vertex sparse graph, where $\log(N) = \text{poly}(m)$. Childs (2009) noted however that in contrast with Feynman whose motivation for constructing the Hamiltonian was to give a physically reasonable description of a computing device, in his work, vertices of graphs by which quantum gates are constructed represent basis states not physical objects such as qubits. Thus the construction does not directly give an architecture for a physical device.

# Chapter 3
# Physical Implementation

There have been numerous proposals for physical implementations of quantum walks utilizing a diverse array of solid state as well as optical schemes. Here we will present a comprehensive survey of the various implementation schemes thus far proposed, and in some cases experimentally demonstrated, grouped in accordance with the physical medium they essentially feature.

Notably, while a number of authors have considered the implementation of *continuous-time* quantum walks, a vast majority of the proposed systems to date have focused on the realization of *discrete-time* or *coined* quantum walks, thus requiring the underlying physical system to

- Exhibit addressable states representing the nodes or position states as well as the sub-nodes or coin states of the walk,
- Allow selective interactions between those states to implement the coin operator, and
- Allow some form of state dependent time evolution to implement the conditional translation operator.

Therefore in considering each coined quantum walk scheme, it is constructive to view the essential physics in the context of the above criteria which, irrespective of the medium, form a common thread among many of the proposals described here.

A concept that will become important in our review of various implementation schemes pertains to the role of quantum phenomena in the outcome of the walk. As noted by a number of authors, despite its name, implementing the single-particle quantum walk does not in fact necessitate the use of an inherently quantum system. As we will see, experimental efforts utilizing classical optics for example, have indeed proven successful in reproducing the signature distribution typically associated with a quantum walk. This highlights the notion that the dynamics of single-particle quantum walks depend only upon the interference of waves evolving in a discrete state space; something that is not a uniquely quantum effect but can in fact be found in many classical systems.

What, if any, is then the contribution of the quantum effects? The answer lies in the efficient provision of computational resources due to quantum entanglement.

J. Wang and K. Manouchehri, *Physical Implementation of Quantum Walks*,
Quantum Science and Technology, DOI 10.1007/978-3-642-36014-5_3,
© Springer-Verlag Berlin Heidelberg 2014

If quantum walks were to be the basis of new computational algorithms, their application to problems of interest would conceivably require a vast number of nodes. This would in turn necessitate utilizing physical systems capable of providing a large number of addressable internal states which may be regarded as computational resources. Whereas in any classical system computational resources scale linearly with the size of the system (or its constituent components), the number of available states in an entangled quantum system scale exponentially. Nonetheless despite its perceived virtues, quantum entanglement only features in a few implementations that employ *quantum circuits*, while a large number of proposed systems whose dynamics are governed by the quantum Hamiltonian of a single quantum walker, do not in fact employ any entanglement and are instead solely reliant on the wave-like interference of probability amplitudes described by the evolution operator.

As described in the concluding section (Sect. 3.9) however, recent theoretical as well as experimental studies of two- and multi-particle quantum walks have revealed rich new dynamics that are observable only in the quantum regime, owning to the non-classical interactions between the participating particles. While presenting new challenges for those concerned with their physical implementation, these inherently quantum extensions of the single-particle walk have the potential to open up sweeping vistas for the development of powerful quantum algorithms.

## 3.1  Linear Optics

In the following proposals basic optical elements, commonly found in most laboratories, are employed to implement quantum walks. These examples clearly demonstrate the notion that implementing a quantum walk does not necessitate the use of an inherently "quantum" system, and that the characteristic quantum walk distribution can be effectively constructed using the interference of classical field.

### 3.1.1  Linear Cavity

An idea explored by many researchers for implementing quantum walks is based on the classic work of Galton (1877) on random walks and a device known as the Galton board or Galton quincunx depicted in Fig. 3.1. Balls are rolling down a sloping board and are scattered by a grid of pins and the random walk performed by the balls leads to a Gaussian diffusion. A variety of schemes were then proposed to design an all optical quincunx, using linear optics elements. The primary difference between the Galton quincunx and its optical counterpart is that an object traversing the Galton quincunx travels along a single path from the source to the detector, whereas each incident object in an optical quincunx can be viewed as simultaneously traversing all possible paths to the detector. This behavior provides a natural basis for implementing the quantum walk.

**Fig. 3.1** Schematics of a
Galton board

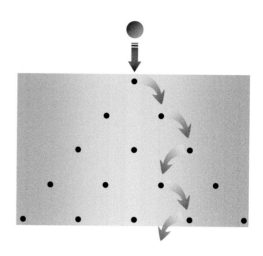

Knight et al. (2003b) described an experiment by Bouwmeester et al. (1999), in the context of the optical quincunx, which in principal implements a quantum walk using the interference of classical field in an optical cavity, without the authors explicitly noting this. In the implementation of Bouwmeester et al. (1999) as depicted in Fig. 3.2, frequency levels inside an optical resonator mimic the rolling balls and birefringent crystals inside the resonator perform the role of scattering pins. Figure 3.3 illustrates the building blocks of the optical quincunx. The linear optical resonator or cavity has equidistant longitudinal modes $m = 0, \pm 1, \pm 2 \ldots$ which represent the eigenstates of the light inside the resonator. The modes are well resolved with angular frequency spacing $\omega_{FSR} = 2\pi f_{FSR}$, where $f_{FSR} = c/2L$, $c$ is the speed of light, and $L$ is the length of the linear resonator. The total light field in the cavity is given by Eq. A.6 where there is only a single spatial mode $\mathbf{k}$ and the direction of propagation is taken to be $z$. Hence

$$\mathbf{E}(z, t) = \sum_{m=-l}^{l} \sum_{\lambda=1}^{2} E_{m,\lambda}\, \mathbf{u}(\lambda)\, e^{i(kz - \omega_m t)} + \text{c.c.}$$

$$= \sum_{m=-l}^{l} \mathbf{E}_m(t) e^{ikz} + \text{c.c.}, \tag{3.1}$$

where we have used subscript $m$ (rather than $\nu$) to label the frequency modes, $\omega_m = \omega_0 + m\omega_{FSR}$ is the angular frequency of the $m$th mode, $\omega_0$ is the carrier frequency, $\mathbf{E}_m(t) = \mathbf{J}\, e^{-i\omega_m t}$, and $\mathbf{J} = (\tilde{E}_x, \tilde{E}_y)$ is the Jones vector, with complex-valued components of the polarization $\tilde{E}_x$ and $\tilde{E}_y$ conveniently taken to lie in the $x$-$y$ plane.

Using the Jones-matrix formalism, the action of each polarization changing optical element inside the resonator can be represented as a $2 \times 2$ matrix. More specifically, the electro-optic modulator $EOM_1$ is represented by

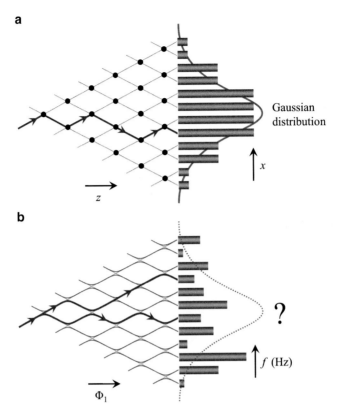

**Fig. 3.2** (**a**) The classical Galton board, sloping downwards in the $z$ direction, yields a Gaussian distribution function for the final position of the particle along the $x$-axis. The *dots* represent pins at which the balls are scattered. (**b**) The grid of Landau–Zener crossings which is the quantum- or wave-mechanical analog of the classical Galton board; $f$ indicates the frequency axis, and $\phi_1$ is a control parameter which is proportional to time (Adapted from Bouwmeester et al. (1999))

$$B_1(t) = \begin{pmatrix} e^{i\phi_1(t)/2} & 0 \\ 0 & e^{-i\phi_1(t)/2} \end{pmatrix}, \tag{3.2}$$

and the electro-optic modulator EOM$_2$, having been rotated over 45° with respect to the optical axis of EOM$_1$, is represented by

$$B_2(t) = \begin{pmatrix} \cos(\phi_2(t)/2) & -i\,\sin(\phi_2(t)/2) \\ -i\,\sin(\phi_2(t)/2) & \cos(\phi_2(t)/2) \end{pmatrix}, \tag{3.3}$$

where $\phi_1(t)$ and $\phi_2(t)$ are the phase values that the two orthogonal polarizations along the axis of refringence obtain by passing EOM$_1$ and EOM$_2$. In their experiment Bouwmeester et al. (1999) linearly increased the voltage $V_1$ across EOM$_1$, making $\phi_1 = \overline{\omega}t$ for a constant $\overline{\omega}$, while keeping the voltage $V_2$ across EOM$_2$ constant, making $\phi_2 = \Delta$ for a fixed $\Delta$. Taking some point between EOM$_1$

**Fig. 3.3** (a) Sketch of a linear optical resonator which has equidistant longitudinal modes ($m = 0, \pm 1, \pm 2$). (b) Including an electro-optic modulator (EOM$_1$) inside the resonator, and increasing the voltage $V_1$ across the modulator, leads to crossing levels with orthogonal polarizations $x$ and $y$. (c) Including a second modulator (EOM$_2$) inside the resonator, rotated over 45° with respected to the optical axis of EOM$_1$ and with a constant applied voltage $V_2$, turns each level crossing into an avoided or Landau-Zener crossing (Adapted from Bouwmeester et al. (1999))

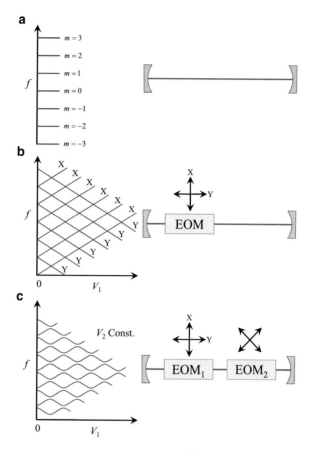

and EOM$_2$ as the starting point and neglecting any optical losses, the state of the light field after a round trip is given by

$$\mathbf{E}_m(t + T) = M(t) \, \mathbf{E}_m(t), \tag{3.4}$$

where $M(t) = B_2(t)^2 B_1(t)^2$ is the round trip matrix and the birefringence of the EOMs per round trip is much smaller than $2\pi$, keeping the round trip matrix approximately constant during a single round trip time $T$. The presence of each EOM inside the resonator modifies the system's eigenvalues and hence the structure of the cavity modes, as depicted in Fig. 3.3. Considering the product $B_1(t)^2 \, \mathbf{E}_m(t)$ for example, it becomes evident that EOM$_1$ increases the frequency associated with the $\tilde{E}_x$ polarization component by $\overline{\omega}$ while decreasing the frequency associated with the $\tilde{E}_y$ polarization component by the same amount after every round trip. This produces the mode splitting grid depicted in Fig. 3.3b with level cross points at $\overline{\omega} = n \, \omega_{\text{FSR}}$ for $n = 1, 2, \ldots$. Similarly, Bouwmeester et al. (1999) showed that the addition of EOM$_2$ turns each crossing into an avoided crossing depicted in Fig. 3.3c.

A closer look at the above formalism however reveals how accurately it describes a quantum walk on a line. Here the role of the walker is played by the light field frequency (longitudinal cavity modes $m$), and the role of the coin is played by its polarization state. To formally make this connection, Knight et al. (2003b) used an abstraction, first introduced by Spreeuw (1998) who constructed a complete classical wave-optics analogy of quantum-information processing (Spreeuw 2001). In this analogy the state of the classical wave (see Appendix A.1 for more details) is described using an abstract state (with a special $|$ $)$ notation)

$$|\psi) = \sum_{\mathbf{k},m,\lambda} A_{\mathbf{k},m,\lambda} |\mathbf{k}, m, \lambda), \tag{3.5}$$

where

$$A_{\mathbf{k},m,\lambda} = \frac{E_{\mathbf{k},m,\lambda}}{\|E_{\mathbf{k},m,\lambda}\|}, \tag{3.6}$$

making

$$\sum_{\mathbf{k},m,\lambda} |A_{\mathbf{k},m,\lambda}|^2 = 1, \tag{3.7}$$

as expected for any quantum state. For a system like that of Bouwmeester et al. (1999) where there is only a single spatial mode $\mathbf{k}$ and polarization modes $\lambda = x, y$, the above state may be reduced to

$$|\psi) = \sum_m R_m |m, x) + L_m |m, y), \tag{3.8}$$

where $L_m = A_{\mathbf{k},m,x}$, $R_m = A_{\mathbf{k},m,y}$ and the basis states $|x)$ and $|y)$ form a polarization "cebit", a term coined by Spreeuw (2001) to describe the classical counterpart of a qubit.

The abstract state $|\psi)$ conveniently describes the complete state of a quantum walk, where the discrete position states can be attributed to frequency states $|m)$ and coin states are given by the two polarization states $|x)$ and $|y)$. In this way a quantum walk can be performed in the usual way via the repeated actions of a coin operator $\hat{C}$ followed by a translation operator $\hat{T}$, i.e.

$$|\psi_n) = \left[\hat{T}\hat{C}\right]^n |\psi_0), \tag{3.9}$$

where $n$ is the number of iterations, $\hat{C}$ performs a general unitary transformation on individual cebits, i.e.

$$\begin{pmatrix} L'_m \\ R'_m \end{pmatrix} = \hat{C} \begin{pmatrix} L_m \\ R_m \end{pmatrix}, \tag{3.10}$$

and $\hat{T}$ does the walking according to

$$\hat{T}|m\rangle|x\rangle = |m+1\rangle|x\rangle, \quad \text{and} \tag{3.11}$$

$$\hat{T}|m\rangle|y\rangle = |m-1\rangle|y\rangle. \tag{3.12}$$

Finally, the intensity of each frequency component of the light field, which is the optical analog of the probability for finding the walker at position $m$ at iteration $n$, is given by

$$P_m(n) = |L_{m,n}|^2 + |R_{m,n}|^2. \tag{3.13}$$

Now reconsidering the experiment of Bouwmeester et al. (1999), if the frequency jumps were made equal to the frequency spacing between the cavity modes $m$, i.e. $\bar{\omega} = \omega_{\text{FSR}}$, then after each round trip we would have

$$B_1(t)^2|m\rangle|x\rangle = |m+1\rangle|x\rangle, \quad \text{and} \tag{3.14}$$

$$B_1(t)^2|m\rangle|y\rangle = |m-1\rangle|y\rangle \tag{3.15}$$

which is simply the conditional translation operator, i.e. $\hat{T} \equiv B_1(t)^2$. Similarly $B_2(t)^2$ is a constant complex valued unitary matrix which represents the coin operator, i.e. $\hat{C} \equiv B_2(t)^2$. Hence each step of the quantum walk is conveniently embodied in a single round trip according to $\hat{T}\hat{C} \equiv M(t)$.

In their experimental setup Bouwmeester et al. (1999) placed one mirror on a piezo element in order to tune one of the longitudinal resonator modes (by definition this is the mode $m = 0$) to the frequency of the linearly polarized single-frequency He-Ne injection laser ($\lambda = 633$ nm). The total resonator length was approximately 100 m which yielded a mode spacing of $\omega_{\text{FSR}} = 9.55\,\text{s}^{-1}$ and a round trip time $T$ of 0.7 μs. The cavity decay time, immediately after the injection of light, was measured to be about 70 μs, the time window in which to perform the actual experiment. The linear voltage ramp applied across EOM$_1$ produced a sweep rate of typically $\bar{\omega} = 3 \times 10^6\,\text{s}^{-1}$, while the voltage applied across EOM$_2$ was kept constant corresponding typically to $\Delta = 0.2\pi$. After a certain number of steps into the grid (i.e. going though a number avoided cross points) the voltage ramp across EOM$_1$ was stopped which landed the system in a stationary state enabling the researchers to analyze over which levels the light was distributed. Figure 3.4 shows the experimental measurements of the spectral diffusion by analyzing the small fraction of light leaking out of the resonator through one of the mirrors. The quantum walk signature is immediately apparent. Nonetheless there is a subtle but essential difference between the experiment of Bouwmeester et al. (1999) and an actual implementation of the quantum walk. Whereas adherence to the exact quantum walk procedure would require a frequency displacement $\bar{\omega} = \omega_{\text{FSR}}$ at every round trip followed by the application of the unitary transformation $B_2(t)^2$, in their experiment the frequency displacement $\bar{\omega}$ was only a fraction of the $\omega_{\text{FSR}}$ while the unitary transformation $B_2(t)^2$ was still applied at every round trip. Strictly speaking then, although providing a proof of principle demonstration, this experiment cannot be regarded as an exact implementation of the quantum walk.

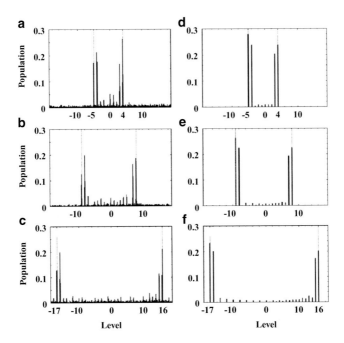

**Fig. 3.4** Experimental results for the spectral diffusion of the laser in the linear cavity. The system was initialized by populating the cavity mode $m = 0$ and parameters $\bar{\omega} = 3 \times 10^6\,\mathrm{s}^{-1}$ and $\Delta = (0.2 \pm 0.02)\pi$. The spectral distributions after (**a**) 8, (**b**) 16, and (**c**) 32 columns of crossings are shown. Graphs (**d**)–(**f**) show the corresponding numerical results, based on the treatment of the optical Galton board as an array of point like beam splitters in the spectral domain. The *dashed lines* indicate the outermost levels which can be populated (From Bouwmeester et al. (1999))

Another difference noted by Knight et al. (2003b) was the dephasing operation $B_2(t)^2$ due to EOM$_2$ which, although unitary, does not correspond to a Hadamard transformation, typically used as the quantum coin in many quantum walk implementations. Hence the authors proposed replacing the EOM$_2$ with a quarter-wave plate (QWP) (see Appendix D.4) with its fast axis forming an angle $\theta = 22.5°$ with respect to the $x$ polarization axis depicted in Fig. 3.5a. Notice that when a QWP is double passed by a mirror reflection, it acts as a half-wave plate (HWP) and rotates the plane of polarization by $2\theta$. A round trip through the QWP would then produce the desired Hadamard transformation. Most importantly Knight et al. (2003b) noted that in many other classical (interferometric) implementations of the quantum walk, the number of necessary optical elements grows quickly with the number of steps in the walk, something that does not occur in this scheme.

### 3.1.2  Optical Rings

In a related work, Knight et al. (2003a) extended the above implementation and proposed a series of modified configurations using a ring cavity illustrated in

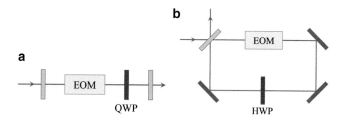

**Fig. 3.5** Schemes for the optical implementation of the quantum walk on a line using polarization cebits. In (**a**), a linear Fabry-Pérot cavity, the electro-optic modulator (EOM) shifts the field frequency up or down in $\overline{\omega}/2$ depending on its polarization, and a quarter-wave plate (QWP) with its axis forming an angle 22.5°, with respect to the $x$-axis, performs the Hadamard transformation (notice that light passes twice through each intra-cavity element every round trip). In (**b**), a ring cavity, the electro-optic modulator (EOM) shifts the field frequency up or down in $\overline{\omega}$ depending on its polarization, and a half-wave plate (HWP) with its axis forming an angle 22.5° with respect to the $x$-axis, performs the Hadamard transformation (notice that light passes only once through each intra-cavity element every round trip) (Adapted from Knight et al. (2003a))

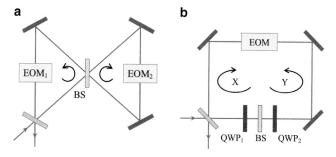

**Fig. 3.6** Schemes for the optical implementation of the quantum walk using position cebits. In (**a**), two coupled unidirectional ring cavities, $EOM_1$ and $EOM_2$, increase and decrease, respectively, the field frequency every round trip and the Hadamard transformation is performed by the beam splitter BS. In (**b**), a bidirectional ring cavity is designed for sustaining the $x$- and $y$-polarized fields in the clockwise and counterclockwise directions respectively. The EOMs increase and decrease, respectively, the $x$- and $y$-polarized field frequencies and the set formed by the two quarter-wave plates $QWP_1$, $QWP_2$, and BS performs the Hadamard transformation (Adapted from Knight et al. (2003a))

Figs. 3.5b, 3.6a and 3.6b. The first configuration (Fig. 3.5b) is a simple modification of the original design presented in Fig. 3.5a, where the light is now traversing each optical element only once in a round trip, and hence the QWP is replaced by a HWP and the EOM shifts the field frequency up or down by $\overline{\omega}$ instead of $\overline{\omega}/2$.

The authors also noted that the frequency shift introduced by the EOM cannot be better resolved than the inverse of the cavity round trip time $\tau_c$, which is precisely $\omega_{FSR}$. Hence the steps of the quantum walk would not be well resolved if $\overline{\omega} = \omega_{FSR}$ as was previously suggested. The authors then proposed a number of ways to overcome this difficulty. One way is to employ frequency jumps several times larger than the free spectral range, i.e. $\overline{\omega} = k\omega_{FSR}$ with $k$ an integer larger than one. In this way the uncertainty in the frequency displacement does not limit the resolution of

the quantum walk steps. Another possibility is to let the frequency shift introduced
by the EOM be smaller than $\omega_{FSR}$. In this case it would take several round trips to
perform a single frequency step of the quantum walk and then it would be necessary
to control the action of the HWP (or the QWP in the linear cavity) that performs the
Hadamard transformation as it should not act until the frequency step of the quantum
walk is completed. For example, if the frequency step takes five cavity round trips,
the HWP should act only once every five round trips. This can be accomplished by
substituting the HWP by a second EOM to which a constant voltage of appropriate
magnitude is applied (similar to the original setup by Bouwmeester et al. (1999)),
only every five cavity round trips.

In the ring cavity design in Fig. 3.6a, Knight et al. (2003a) introduced an
alternative coin space where instead of using a polarization cebit, they used a
position cebit characterized by the two different paths of light within the cavity,
labeled by $|\mathbf{k}_1\rangle$ and $|\mathbf{k}_2\rangle$. Hence the complete state of the quantum walk is given by

$$
\begin{aligned}
|\psi\rangle &= \sum_{\mathbf{k},m,\lambda} A_{\mathbf{k},m,\lambda} |\mathbf{k},m,\lambda\rangle \\
&= \sum_{\mathbf{k},m} A'_{\mathbf{k},m} |\mathbf{k},m\rangle (c_x|x\rangle + c_y|y\rangle) \\
&= \sum_{\mathbf{k},m} A'_{\mathbf{k},m} |\mathbf{k},m\rangle |p\rangle \\
&= \sum_{m} R_m |m,\mathbf{k}_1\rangle + L_m |m,\mathbf{k}_2\rangle,
\end{aligned}
\tag{3.16}
$$

where $|p\rangle$ is the polarization state of radiation which remains unchanged throughout
the experiment and is therefore left implicit in the last line, $L_m = A'_{\mathbf{k}_1,m}$ and $R_m =$
$A'_{\mathbf{k}_2,m}$. Two electro-optic modulators, EOM$_1$ and EOM$_2$, are needed here to operate
on either sides of the cavity, which now represent the coin states. The action of
these EOMs must be polarization independent. Alternately, the polarization of light
and the axes of both EOMs must be aligned. One of the EOMs increases the field
frequency by $\overline{\omega}$, while the other decreases it by the same amount. The Hadamard
transformation is performed by the beam splitter (BS) and additional phase shifters
(Spreeuw 2001) which are not represented in the figure. In fact any unitary coin
transformation of the type $U(2)$ can be implemented using a lossless beam splitter
and a phase shifter at one of the output ports (Cerf et al. 1997). Knight et al. (2003a)
also noted that in this implementation one can consider more than two optical paths,
and thus coin-cebits with more than two components, which can be useful for the
implementation of multidimensional quantum walks.

In Fig. 3.6b, a bidirectional ring cavity, the horizontal and vertical polarization
components of the field are forced to travel in counter propagating directions. With
a $y$-polarized input field initially entering the cavity from the left, the lineup of
the two quarter-wave plates, QWP$_1$ with axis oriented at $45°$ and QWP$_2$ with axis
oriented at $-45°$, and the beam splitter guarantees that the clockwise field remains
$x$-polarized and the counterclockwise field remains $y$-polarized. This is because

when traversing the three elements, the polarization of a beam does not change, since the effect of QWP$_1$ is canceled out by the effect of QWP$_2$, and vice versa. For light reflected by the BS, on the other hand, QWP$_1$ or QWP$_2$ is crossed twice, which is equivalent to the effect of a HWP with axis at 45° and −45° respectively. Hence QWP$_1$ changes the polarization of the reflected light from $y$ to $x$ while QWP$_2$ changes the polarization of the reflected light from $x$ to $y$. The quantum walk implemented by this design is based on a hybrid between polarization and direction cebits, where $|x, \mathbf{k}_1\rangle$ and $|y, \mathbf{k}_2\rangle$ are the two coin states with $\mathbf{k}_1$ and $\mathbf{k}_2$ denoting the two propagation directions. The complete state of the walk is then given by

$$|\psi\rangle = \sum_{\mathbf{k},m,\lambda} A_{\mathbf{k},m,\lambda} |\mathbf{k}, m, \lambda\rangle$$

$$= \sum_m R_m |m, x, \mathbf{k}_1\rangle + L_m |m, y, \mathbf{k}_2\rangle, \qquad (3.17)$$

where $L_m = A_{\mathbf{k}_1,m,x}$ and $R_m = A_{\mathbf{k}_2,m,y}$ are the normalized radiation amplitudes entering the beam splitter from either direction. A general unitary coin operation $\hat{C}$ is performed by the beam splitter (and additional phase shifters not shown in the figure) acting on the spatial cebit, which by design corresponds directly to the polarization cebit. The polarization cebit on the other hand is used to couple the quantum walk with the EOM which performs the translation operation $\hat{T}$, by increasing and decreasing the field frequency, in $\overline{\omega}$ units, for the $x$ and $y$ polarization components respectively. Hence

$$\hat{T}|m\rangle|x, \mathbf{k}_1\rangle = |m + 1\rangle|x, \mathbf{k}_1\rangle, \quad \text{and} \qquad (3.18)$$

$$\hat{T}|m\rangle|y, \mathbf{k}_2\rangle = |m - 1\rangle|y, \mathbf{k}_2\rangle. \qquad (3.19)$$

The authors noted that in the two schemes shown in Fig. 3.6, the output corresponds only to one of the two cebit states $P_m^R = |R_m|^2$ or $P_m^L = |L_m|^2$. In order to obtain the complete quantum walk distribution (see Eq. 3.13), one should instead allow the light to exit the cavity through two of the cavity mirrors and then combine the two beams into a single beam.

Finally, Knight et al. (2003a) presented a scheme for modifying the above designs in order to implement a quantum walk on a circle. If $2M - 1$ is the number of possible discrete values for the frequency, this scheme provides a way of shifting the frequency $\omega + (M + 1)\overline{\omega}$ to $\omega - M\overline{\omega}$ and the frequency $\omega - (M + 1)\overline{\omega}$ to $\omega + M\overline{\omega}$. In the scheme of Fig. 3.6a for example, this is done by substituting the EOMs with the device shown in Fig. 3.7, denoted by $\overline{\text{EOM}}$. It consists of two electro-optic modulators and two specially designed mirrors. Mirrors M1 and M2 only reflect the frequencies $\omega + (M + 1)\overline{\omega}$ and $\omega - (M + 1)\overline{\omega}$, respectively, and are transparent to all other frequencies in the spectrum. Consider any step at which the field entering $\overline{\text{EOM}}$ contains all the allowed frequencies from $\omega - (M + 1)\overline{\omega}$ through to $\omega + (M + 1)\overline{\omega}$. The role of M1 is to separate the frequency $\omega + (M + 1)\overline{\omega}$, which is directed to EOMa, from the rest of frequencies, which are directed to EOMb. The modulators are configured such that EOMa decreases the entering

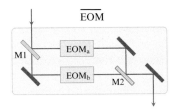

**Fig. 3.7** Scheme of the $\overline{\text{EOM}}$ device that replaces the electro-optic modulators in Fig. 3.6a in order to perform the quantum walk on a circle. M1 and M2 are mirrors that reflect a single frequency ($M\overline{\omega}$ and $-M\overline{\omega}$, respectively) and EOMa and EOMb are electro-optic modulators (Adapted from Knight et al. (2003a))

frequency in $2M\overline{\omega}$, while EOMb increases the frequencies entering it in $\overline{\omega}$. After traversing the electro-optic modulators, the frequencies enter M2. The frequencies $\{\omega - (M-1)\overline{\omega}, \ldots, \omega + M\overline{\omega}\}$ that come from EOMb traverse M2, while the frequency $\omega - M\overline{\omega}$ that comes from EOMa is reflected by M2. Then, the set of frequencies exiting $\overline{\text{EOM}}$ is the same set that entered, but all of them have been shifted appropriately, i.e. all frequencies are increased by $\overline{\omega}$ except $\omega + M\overline{\omega}$ which is converted into $\omega - M\overline{\omega}$. One can obtain an inverse $\overline{\text{EOM}}$ in which all frequencies are decreased by $\overline{\omega}$ except $\omega - M\overline{\omega}$ which is converted into $\omega + M\overline{\omega}$. This is done by interchanging the positions of M2 and M1 and then modifying EOMa and EOMb such that EOMa now increases the incoming frequency in $2M\overline{\omega}$ whilst EOMb decreases in $\overline{\omega}$ the rest of frequencies. The authors suggested that the device $\overline{\text{EOM}}$ could also be used in the cavities of Figs. 3.5b and 3.6b by suitably adapting its operation.

Bañuls et al. (2006) extended one of Knight et al.'s implementations (2003a), depicted in Fig. 3.5b, to incorporate a time-dependant coin leading to quantum walks which exhibit interesting localization and quasi-periodic dynamics. The time-dependant coin is given by $\hat{C}(t) = \hat{C}_0(t)\hat{C}$, where $\hat{C}$ is an arbitrary constant coin matrix and

$$\hat{C}_0(t) = \begin{pmatrix} e^{-i\phi_0 t} & 0 \\ 0 & e^{i\phi_0 t} \end{pmatrix}, \tag{3.20}$$

for some constant phase $\phi_0$. Figure 3.8 illustrates the proposed setup for implementing this quantum walk. It is essentially the same setup as the one proposed by Knight et al. (2003a), but with an additional EOM element (labeled EOMbis) placed between the original EOM and the wave plate. Without the EOMbis the setup implements an ordinary quantum walk with a constant coin operator $\hat{C}$ as described by Knight et al. (2003a). Implementing the operator $\hat{C}_0(t)$ for a given $\phi_0$ necessitates the addition (subtraction) of $\phi_0 t$ to the phase of the horizontal (vertical) polarization component of the field. This is easily carried out by introducing the EOMbis to which a suitable constant voltage is applied. Now, in order to implement $\hat{C}(t)$, this added (subtracted) phase must be increased at each cavity round trip, which is done by applying a staircase voltage to EOMbis as represented in Fig. 3.8. The voltage must remain constant while the light pulse is traversing EOMbis, in order to modify

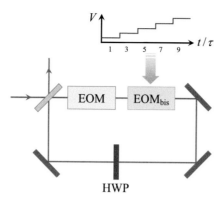

**Fig. 3.8** The proposed experimental setup for implementing the quantum walk with a time-dependent coin. The *upper part* represents the staircase voltage that has to be applied to the EOMbis in order to implement the time-dependent coin, and $\tau$ is the cavity round trip time (Adapted from Bañuls et al. (2006))

the phase and not the field frequency, and then be rapidly increased for the phase increment to take the value $\phi_0(t + 1)$ in the subsequent round trip.

A distinguishing feature of implementation schemes such as the cavity ring configurations proposed by Knight et al. (2003a) is the use of an optical feedback loop which provides a key advantage: the amount of required resources remains constant as the number of quantum walk steps (round trips in the loop) increases. Noting this advantage, Schreiber et al. (2010) experimentally demonstrated a robust scheme utilizing an optical feedback loop, but in this case employing only passive optical elements. Akin to a number of earlier schemes, the coin space of the walk in this implementation is spanned by the linear polarization states $|H\rangle$ and $|V\rangle$ of the input laser. But in a novel approach the authors demonstrated the encoding of position states as temporal information carried by single photons.

A sketch of this experimental setup is presented in Fig. 3.9. The quantum walker is an 88 ps laser pulse, produced by an 805 nm source with repetition rate of 1 MHz and attenuated to the single-photon level by using neutral density filters. The state of the walker is initialized using a polarizing beam splitter together with standard half- and quarter-wave plates. It is then injected into the optical loop using a 50–50 beam splitter. At each step of the walk corresponding to a round trip, there is a 50 % probability of coupling the photon out of the loop, in which case an avalanche photodiode (APD) with time jitter <1 ns will register a click.

The implementation of the coin operation is straight forward using a half-wave plate with its fast axis at an angle $\theta$ with respect to the horizontal. The resulting coin operator is given by

$$\hat{C} = \begin{pmatrix} \cos(2\theta) & \sin(2\theta) \\ \sin(2\theta) & -\cos(2\theta) \end{pmatrix}, \tag{3.21}$$

as described in Appendix D.4.

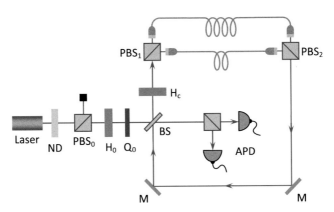

**Fig. 3.9** Schematic diagram of the setup used to perform the quantum walk on a line. A laser field is attenuated to the single-photon level via neutral density filters (ND) and coupled into the network loop through a 50–50 beam splitter (BS). A polarizing beam splitter $PBS_0$, half-wave plate $H_0$, and quarter-wave plate $Q_0$ are utilized to prepare the quantum walker's initial state. The coin half-wave plate $H_c$ and polarizing beam splitters $PBS_1$ and $PBS_2$ perform the walk. Mirrors M close the network loop. Measurements are performed using avalanche photodiodes (APD) (Adapted from Schreiber et al. (2010, 2011))

   To perform the conditional translation, Schreiber et al. (2010) employed two polarizing beam splitters $PBS_1$ and $PBS_2$ connected by a pair of polarization-maintaining single mode optic fibers. The horizontal and vertical components of the input light are first spatially separated by $PBS_1$. Each component then travels along one of the optic fibers before being recombined at the output of $PBS_2$. Hence by carefully adjusting the length of each fiber, it is possible to introduce an additional temporal separation between the two pulse components which persists after recombination. Considering the state of a photon entering $PBS_1$ at time $t = 0$, its passage through the fiber network F can be described by the transformation

$$|H, 0\rangle \xrightarrow{\text{F}} |H, T_0 + \delta t\rangle,$$

$$|V, 0\rangle \xrightarrow{\text{F}} |V, T_0 - \delta t\rangle, \tag{3.22}$$

where $T_0 = (T_H + T_V)/2$ and $\delta t = |T_H - T_V|/2$ with $T_H = 40\,\text{ns}$ and $T_V = 45\,\text{ns}$ representing the travel time of each component through its carrier fiber. Using this formalism, position states of the walk at the $N$th round trip can now be encoded as $|t_k\rangle_N \equiv |NT_0 + k\delta t\rangle$ for $k = \pm 0, 1, 2 \ldots N$, corresponding to the range of possible times at which the output photon may be detected by the photo detectors. Here $\delta t$ represents the quantum walk step length and we may drop the global constant $NT_0$ which simply represents the median detection time. Using this definition to rewrite the fiber network transformation we obtain

$$|H, t_k\rangle \xrightarrow{\text{F}} |H, t_{k+1}\rangle,$$

$$|V, t_k\rangle \xrightarrow{\text{F}} |V, t_{k-1}\rangle, \tag{3.23}$$

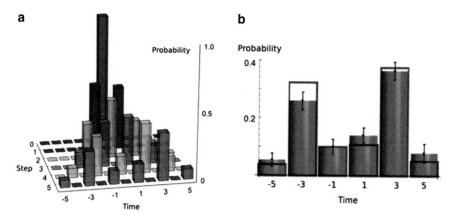

**Fig. 3.10** Measured probability distribution of the photon's arrival time. (**a**) Evolution of the distribution from the initial circularly polarized state (*rear part*) to the state after the fifth step (*front part*). (**b**) Detail of the measured distribution after five steps. *Filled bars*: measured results. *Frames*: predictions from the theoretical model (From Schreiber et al. (2010))

which is readily recognizable as the conditional translation, and the entire position space has undergone an additional but unimportant global shift $T_0$.

Constructing the final quantum walk probability distribution involves a series of consecutive runs of the experiment, each generating at most a single click at a specific time, which is then recorded by a computer via a time-to-digital converter interface. Using this apparatus Schreiber et al. (2010) performed a five step quantum walk, with the walker's initial coin state represented by a circular polarization state $|H\rangle + i|V\rangle$ and the coin half-wave plate angle set to $\theta = 22.5°$ to perform the Hadamard rotation. The resulting probability distribution, carrying the usual quantum walk signature, is shown in Fig. 3.10. As noted by the authors, phase stability is required only during the short time scale (225 ns) of a single experiment, in contrast to the longer time required for an ensemble measurement. Consequently no active phase stabilization was used in the experiment.

Schreiber et al. (2010) also performed a detailed theoretical analysis of the system by taking into account possible sources of coherent and incoherent errors. These factors were combined to produce an effective coin operator

$$\hat{C} = L(\epsilon_F)R(\varphi)R(\theta)L(-\epsilon_{HWP})R(-\theta)L(\epsilon_{BS}), \tag{3.24}$$

where

$$L(\epsilon) = \begin{pmatrix} 1 & 0 \\ 0 & \epsilon \end{pmatrix}, \tag{3.25}$$

is a matrix characterizing differential losses, $\epsilon$ is the efficiency ratio with values $<1$ indicating loss imbalance between the $|H\rangle$ and $|V\rangle$ polarization states, and

$$R(\alpha) = \begin{pmatrix} \cos(\alpha) & \sin(\alpha) \\ \sin(\alpha) & -\cos(\alpha) \end{pmatrix}, \tag{3.26}$$

is the usual rotation matrix. Characterizing the error contribution due to various components, the authors determined $\epsilon_{BS} = 0.99$ for the coupling beam splitter, $\epsilon_F = 0.96$ at the fiber network, $\epsilon_{HWP} = 0.98$ between the slow and the fast axis of the coin half-wave plate and an undesired polarization rotation $\varphi = 1.4°$ introduced by the mirrors. The results indicate that compared to photons in $|V\rangle$ polarization state, those in $|H\rangle$ state experience a greater loss, experimentally measured to be 3 % per step. This is evident in the form of asymmetry characterizing the probability distributions if Fig. 3.10.

Schreiber et al. (2010) suggested that using the best available optical components, the setup efficiency would reasonably increase from its current experimental value of 18 % to as high as 71 %. Furthermore the authors estimated that by replacing the single photon input with a 1 W laser pulse (see Jeong et al. (2004) for the relationship between the coherent and single photon input) and the addition of an active switch to couple the photon out of the loop, the signal-to-noise ratio could be further improved, thus allowing the walk to reach up to 100 steps.

In a subsequent work Schreiber et al. (2011) did in fact improved on the efficiently of their setup raising it to 55 %, which allowed them to increase the number of viable steps to 28. Moreover, the authors demonstrated the implementation of a tunable coin operator by introducing an electro-optic modulator (EOM) in the optical loop, placed between the coin half-wave plate and PBS$_1$ in Fig. 3.9. This however resulted in additional losses which reduced the overall efficiency (down to 22 %) and the number of possible steps. Passage through the EOM introduces a tunable phase shift $\phi_{V/H}$ in each component of polarization resulting in a modified coin operator given by

$$\hat{C} = \begin{pmatrix} e^{i\phi_H} & 0 \\ 0 & e^{i\phi_V} \end{pmatrix} \begin{pmatrix} \cos(2\theta) & \sin(2\theta) \\ \sin(2\theta) & -\cos(2\theta) \end{pmatrix}, \tag{3.27}$$

subject to the condition $\phi_V/\phi_H \approx 3.5$ imposed by the EOM properties. Changing the parameters $\phi_{H/V}$ and $\theta$ in a controlled way makes it possible to alter the transition coefficients in Eq. 1.58 and hence create a diverse range of physical conditions for the quantum walk evolution. Schreiber et al. (2011) used this tunable coin to implement a single particle quantum walk in an environment with (a) static and (b) dynamic disorder.

To implement a quantum walk with static disorder the coin operation $\hat{C}(x)$ is required to be position and not step dependent. The authors realized this by applying a carefully adjusted periodic noise signal (with intervals $\delta t$) to the EOM, generating a random but static sequence of phase shifts $\phi_{H/V}(x)$ with $\phi_V(x) \in [-\Phi_{max}, \Phi_{max}]$, such that the photon acquires the same phase any instance it appears at position $x$. Figure 3.11a shows the resulting probability distribution after 11 steps with $\Phi_{max} = (1.14 \pm 0.05)\pi$, where different phase patterns at subsequent runs have been used to average over various disorders, as considered in the model of Anderson (1958).

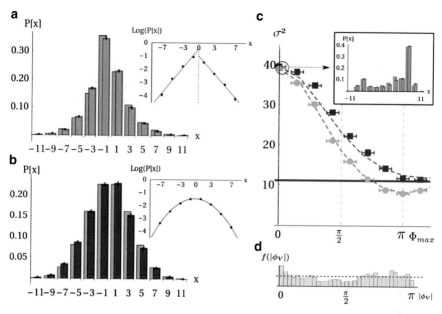

**Fig. 3.11** Measured probability distribution (*front*) and respective theory (*back, gray bars*) of 11 steps of a quantum walk ($\theta = 8$) with static disorder (**a**), dynamic disorder (**b**), and in a decoherence-free environment [*inset* (**c**)]. The *insets* in (**a**) and (**b**) show the measured distribution in semilog scale with linear (**a**) and parabolic fit (**b**). (**c**) Transition of the variance from ballistic quantum walk to diffusive or localized evolution due to dynamic (*red squares*) and static (*green dots*) disorder with increasing disorder strength $\Phi_{max}$; *dashed lines*: theory with adaption for experimental imperfections. The *red solid line* marks the variance of a classical random walk. (Vertical error is smaller than the dot size.) (**d**) Relative frequency $f(|\phi_V|)$ of the applied phases $\phi_V$ for the signal with interval $\Phi_{max} = (1.02 \pm 0.05)\pi$. The *dashed line* indicates the uniform distribution (From Schreiber et al. (2011))

A striking signature of Anderson localization is emphasized by the linear fits in the semilog scaled plot displayed in the insert.

To generate a system with dynamic disorder, the temporal length of the EOM noise signal is detuned to eliminate position dependent phase correlations. As a result, the photon undergoes a classical random walk, revealing a binomial probability distribution as shown in Fig. 3.11b. Furthermore in contrast to the previous case, the spatial profile of the wave-packet shows a parabolic shape in the semilog scale. As evident in Fig. 3.11c, a stepwise increase of the disorder strength $\Phi_{max}$ demonstrates the controlled transition of the system from the ballistic evolution (decoherence-free quantum walk) towards either a diffusive evolution or localization depending on the type of disorder (dynamic or static) being introduced.

Regensburger et al. (2011) adopted a similar approach to Schreiber et al. (2011) where the light propagates in two single-mode fiber loops of differing lengths, thus emulating the quantum walk position space by time-multiplexing. Crucially however, the authors introduced a semiconductor optical amplifier in each loop with its gain carefully adjusted to compensate only for the signal losses caused by the absorption and detection of light during each round trip. The amplifiers do not affect

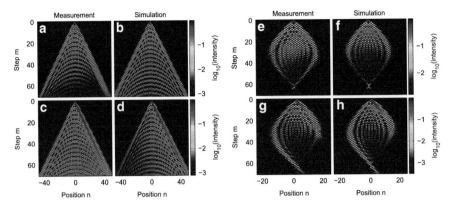

**Fig. 3.12** Evolution of the lossless quantum walk starting in state $|\downarrow,0\rangle$. *Upper* and *lower panels* correspond to intensities $|\langle\psi|\uparrow,n\rangle|^2$ in the *upper* and $|\langle\psi|\downarrow,n\rangle|^2$ in the *lower loop*. (**a**)–(**d**) Measurements and simulations of ballistic spreading without phase modulation ($\alpha = 0$). (**e**)–(**h**) Discrete-time Bloch oscillations for linear phase gradient $\alpha = 2\pi/32$ (From Regensburger et al. (2011))

classical wave interference (see Jeong et al. (2004) for the relationship between the coherent and single photon input) and the system behaves as if it was lossless, allowing for a considerably larger number of steps than previously reported by Schreiber et al. (2011). In this scheme the coin basis states $|\uparrow\rangle$ and $|\downarrow\rangle$ correspond to the light propagating in the upper and lower loops connected via a 50/50 coupler. The effective coin operator for the system is given by

$$\hat{C} = \frac{1}{\sqrt{2}} \begin{pmatrix} 1 & 0 \\ 0 & e^{in\alpha} \end{pmatrix} \begin{pmatrix} 1 & i \\ i & 1 \end{pmatrix}, \tag{3.28}$$

where the quantum walker's position $n = \ldots -2, -1, 0, 1, 2 \ldots$ and the position dependent phase $n\alpha$ is introduced by a phase modulator incorporated in the lower loop. The resulting quantum walk distribution for a walker initially in state $|\downarrow,0\rangle$ is shown in Fig. 3.12. Notably, without phase modulation ($\alpha = 0$), one observes a ballistic spreading of the light field consistent with the characteristic quantum walk of a single particle on a line, and the initial asymmetry which is caused by injecting the pulse only into the lower loop remains conserved (Figs. 3.12c and 3.12d). On the other hand, applying a phase shift ($\alpha \neq 0$), which grows linearly in position, to the $|\downarrow,n\rangle$ states at every propagation step $m$, leads to discrete-time photonic Bloch oscillations; a quasiperiodic dynamic investigated by Wójcik et al. (2004).

### 3.1.3  Michelson Interferometer

More recently Pandey et al. (2011) demonstrated an experimental realisation of a coined quantum walk of light in frequency space utilizing a series of modified

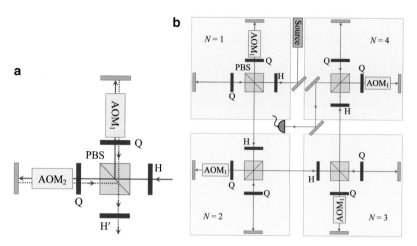

**Fig. 3.13** (**a**) Schematic of a modified Michelson interferometer. *Solid* and *dashed arrows* depict the path of light during its first and second transit through the PBS. (**b**) Experimental setup for a four-step quantum walk. Each *box* constitutes a step of the walk (Adapted from Pandey et al. (2011))

Michelson interferometers. Conceptually, this design employs similar principles to earlier proposals by Knight et al. (2003b), illustrated in Fig. 3.6. A polarizing beam splitter (PBS) ensures that only one component of input light's polarization enters each arm of the interferometer. A pair of acousto-optic modulators (AOMs) in each arm plays the same role as the EOMs in the design of Knight et al. (2003b), shifting the input light frequency up in one arm (AOM$_1$) and down in the other (AOM$_2$). Unlike those earlier proposals however, the coin operation is performed directly on the polarization states (rather than spatial modes) before light enters the interferometer.

Formally, the state of the walker can be appropriately described using Eq. 3.8, where frequency modes $|m\rangle$ and polarization modes $|x\rangle$ and $|y\rangle$ of a coherent input light represent the position and coin states of the walk respectively. Figure 3.13a shows the schematics of the design proposed by the authors that is capable of performing a single step of the quantum walk. The Hadamard coin operation

$$\hat{C} = \frac{1}{\sqrt{2}} \sum_m |m, x+y\rangle(m, x| + |m, x-y\rangle(m, y| \qquad (3.29)$$

is performed utilizing a half-wave plate H, with its fast axis at an angle of 22.5° with respect to the $x$ polarization axis, placed at the entrance of the PBS. The quarter-wave plates Q, with their fast axes at 45° to the $x$ polarization axis, rotate the polarization of light in each arm such that after a double transit it is completely flipped. Hence the beam component initially reflected (transmitted) by the PBS on entry to the interferometer will now be transmitted (reflected), ensuring that both components emerge out of a common PBS face fronting the half-wave plate H′.

With the AOMs in the standard "double-pass" configuration, the combined action of the AOMs and Q plates on the input light can be described by the operator

$$\hat{T}_0 = \sum_m |m + 1, y)(m, x| + |m - 1, x)(m, y|, \qquad (3.30)$$

shifting the frequency modes up and down conditioned on the polarization state of the input light. The half-wave plate H′, with its fast axis at 45° with respect to the $x$ polarization axis, corrects for the unwanted polarization flip experienced by light due to the action of Q plates. Including the action of H′ recovers the standard conditional translation operator

$$\hat{T} = \left( \sum_{m'} |m', x)(m', y| + |m', y)(m', x| \right) \hat{T}_0$$

$$= \sum_m |m + 1, x)(m, x| + |m - 1, y)(m, y|. \qquad (3.31)$$

In order to reduce the number of optical components however, Pandey et al. (2011) removed AOM$_2$ and H′ from their experimental setup, noting that the probability distribution of the resulting quantum walk due to the modified conditional translation operator

$$\hat{T}_{\text{modified}} = \sum_m |m + 1, y)(m, x| + |m, x)(m, y|, \qquad (3.32)$$

would be essentially the same as that of a standard walk, up to a shift and squeezing of the frequency scale. Figure 3.13b depicts the experimental setup used by Pandey et al. (2011) to perform the first four steps of a the quantum walk.

Using this setup, Pandey et al. (2011) also investigated the effect of introducing decoherence via controlled dephasing of the RF input applied to AOMs. By adding white noise with the desired amplitude, the authors were able to demonstrate a transition of the walk's behaviour from quantum to classical over the four steps. Highlighting the possibility of independently controlling the phase shift applied to each AOM, Pandey et al. (2011) suggested that one could in principle extend the present experiment to create time-dependent coins and time-dependent walk steps as well as anisotropic walks.

Lastly, Pandey et al. (2011) noted that the number of optical components for their proposed setup increases linearly with the number of steps in the walk. To remedy this disadvantage the authors proposed a modified implementation scheme using a folded geometry as shown in Fig. 3.14. The additional mirrors and the nonpolarizing beam splitter with a high reflectivity would result in photons reaching the detector after $0, 1, 2, \ldots, N$ steps. By careful matching of path lengths and beam alignment, one could also obtain a cavity implementation, leading to higher fidelity for walks with a large number of steps. Nonetheless as noted by the authors, the limited efficiency of AOMs ($\sim$60 % in the double-pass configuration) would be the main source of loss in any such interferometric schemes.

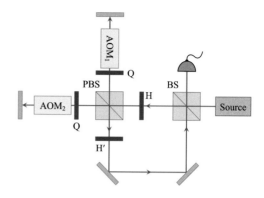

**Fig. 3.14** Multipass modified Michelson interferometer scheme for a compact implementation of the quantum walk of light in frequency space (Adapted from Pandey et al. (2011))

### 3.1.4 Optical Networks

In another approach, Zhao et al. (2002) proposed constructing a network of polarization beam splitters (PBS) and half-wave plates (HWP) to implement the quantum walk. In their proposal the horizontal polarization state $|H\rangle$, and vertical polarization state $|V\rangle$ were used to represent the quantum coin states. When a superposition state passes through a PBS, the PBS transmits the $|H\rangle$ component and reflects the $|V\rangle$ component. Therefore, as depicted in Fig. 3.15, the PBS can be thought of as directing the incoming light into two output ports denoted by "right" and "left" for the $|H\rangle$ and $|V\rangle$ components respectively. However, when a superposition state enters the PBS from the top side ("left" output port of another PBS unit), it results in an incorrect mapping since the $|H\rangle$ component is now transmitted to the "left" port and the $|V\rangle$ component is reflected to the "right" port. Thus the authors introduced a modified PBS (the $\overline{\text{PBS}}$) which transmits the $|V\rangle$ component and reflects the $|H\rangle$ component, using half-wave plates (HWP). They suggested that $\overline{\text{PBS}}$ could be realized by rotating the polarization of the photon by 90° using a HWP with $\theta = 45°$ followed by a rotation back using another HWP on each of the PBS outputs (see $R_{90}$ in Fig. 3.15). After passing through the $\overline{\text{PBS}}$, the $|H\rangle$ component correctly moves to the "right" and the $|V\rangle$ component to the "left". Also, where there are different components incident from both directions, one may conveniently superimpose them using a single PBS as shown in Fig. 3.15. Therefore, utilizing the PBS and the $\overline{\text{PBS}}$ (the modified PBS), the authors were able to define a mechanism for inducing photonic movement depending upon its polarization.

The complete network of optical elements, proposed to implement the quantum walk, is depicted in Fig. 3.16, where $R_{45}$ represents a HWP with $\theta = 22.5°$ and $D$ denotes a photon detector. The similarity with Galton's quincunx is immediately apparent. The authors also defined a *dynamic line* labeled with $j = 0, 1, 2 \ldots$ in Fig. 3.16 representing the nodes of the quantum walk on a line. In this way a single step of the quantum walk involves going from the dynamic line $j$ to the line $j + 1$. By labeling the optical elements along the $j$th dynamic line as $k = -j, -j + 2, \ldots j - 2, j$ we can define the position states of the walk as $|k\rangle$

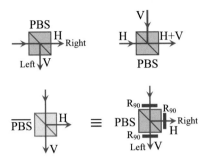

**Fig. 3.15** Schematic of the setup to make use of polarization beam splitter (PBS) and half-wave plate (HWP) as the basic elements for implementing the quantum walk on a line (Adapted from Zhao et al. (2002))

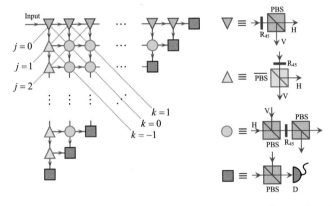

**Fig. 3.16** Schematic of the optical network to implement quantum walk on a line with $N$ steps (Adapted from Zhao et al. (2002))

and attempt to describe the coin and translation operators involved in the quantum walk. Now suppose we are at the $k$th node of the dynamic line $j$ which is receiving its input states from the line $j-1$. Elements marked as a triangle have a single input state, either $|H\rangle$ or $|V\rangle$, and elements marked as a circle have two inputs which are mixed into a single superposition state using a PBS. The input polarization states are then rotated using $R_{45}$ QWP's which represent the coin operator $\hat{C}$ performing a Hadamard transformation. In fact other generalized coin operators can be implemented by an arbitrary choice of the angle $\theta$. Next, PBS and $\overline{\text{PBS}}$ units perform a polarization dependent splitting of the input state which is then passed on to the next dynamic line $j+1$. If the photon is in the $|H\rangle$ polarization state it will be emerging from the "right" output port, which is then automatically incident on the $(k+1)$th unit of the next dynamic line. Likewise if the photon is in the $|V\rangle$ polarization state it will be emerging from the "left" output port, which is then automatically incident on the $(k-1)$th unit of the next dynamic line. This can be conveniently represented by the conditional translation operator $\hat{T}$ such that

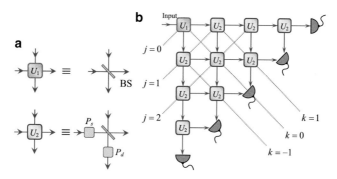

**Fig. 3.17** An all optical setup for the simulation of quantum walks on a line. (**a**) Two different kinds of operations are shown used to construct the optical network. $U_1$ is an ordinary beam splitter BS and $U_2$ comprises a phase shifter $P_s(\omega_1)$, a beam splitter BS and another phase shifter $P_d(\omega_2)$. (**b**) Proposed setup, shown up to the fourth dynamic line. Apart from the input state, all the other modes are initially prepared in vacuum states (Adapted from Jeong et al. (2004))

$$\hat{T}|k\rangle_j|H\rangle = |k+1\rangle_{j+1}|H\rangle \quad \text{and} \tag{3.33}$$

$$\hat{T}|k\rangle_j|V\rangle = |k-1\rangle_{j+1}|V\rangle, \tag{3.34}$$

where $|k\rangle_j$ denotes the $k$th node along the $j$th dynamic line. To implement the quantum walk, a source that produces single photons is required. This can be possible using quantum-dot single photon sources (Kim et al. 1999; Lounis and Moerner 2000; Kurtsiefer et al. 2000; Foden et al. 2000) or with parametric down conversion (Kwiat et al. 1995) by performing a measurement on one photon. A general initial state $|\psi_i\rangle = \cos\theta|H\rangle + e^\phi \sin(\theta)|V\rangle$ can be easily obtained using a set of wave plates (Englert et al. 2001). The single photon is then allowed to pass through the optical network, producing a final state $|\psi_f\rangle$. The final state can be probabilistically measured by positioning detectors behind the PBS's at the final step. The authors pointed out that given the available linear optics technology and the single-photon source (although demanding at the time), their proposed scheme would be at the reach of experimental implementation, be it for a modest number of steps.

Later, Jeong et al. (2004) proposed an optical network similar in structure to the work of Zhao et al. (2002), where the position states of the walk $|k\rangle$ are represented by the nodes along a dynamic line depicted in Fig. 3.17. In this proposal however the nodes are optical units which are constructed using ordinary (polarization insensitive) beam splitters and phase shifters (see Appendix D). Consequently the coin space of the walk is no longer based on the polarization states of light but is rather defined by the "sideward" traveling and the "downward" traveling field modes entering and exiting the optical units.

The input and output states of each optical unit are given by $|A, B\rangle_{input}$ and $|A, B\rangle_{output}$, where $A$ and $B$ represent the state (e.g. Fock state, coherent state, etc.) of the sideward and downward modes respectively. The authors first described the operation of the apparatus when a single photon is used as the input to the network.

In this case the coin states are conveniently defined by $|s\rangle = |1,0\rangle$ (i.e. the state with the photon in the sideward mode) and $|d\rangle = |0,1\rangle$ (i.e. the state with the photon in the downward mode) and the conditional translation operator $\hat{T}$ is trivially embedded in the propagation of the photon from one optical unit to the next. In other words if the photon exits the $k$th unit in the sideward direction it will be incident on the $(k+1)$th unit of the next dynamic line, and if it exits the $k$th unit in the downward direction it will automatically arrive at the $(k-1)$th unit of the next dynamic line. Hence

$$\hat{T}|k\rangle_j|s\rangle = |k+1\rangle_{j+1}|s\rangle \quad \text{and} \tag{3.35}$$

$$\hat{T}|k\rangle_j|d\rangle = |k-1\rangle_{j+1}|d\rangle, \tag{3.36}$$

where $|k\rangle_j$ denotes the $k$th node along the dynamic line $j$ corresponding to the $j$th step of the walk. Each optical unit can be modeled using a simple unitary transformation. The $U_1$ optical unit which consists only of a single beam splitter, is designed to initialize the state of the quantum walk by transforming its single mode input state $|s\rangle$ to a desirable superposition of $|s\rangle$ and $|d\rangle$ states exiting the beam splitter. The $U_2$ optical units include two phase shifters placed at the input and output of the beam splitter which simply allow for a more generalized unitary transformation. This unitary transformation is precisely the coin operator $\hat{C}$ required to mix the input states, prior to the conditional translation $\hat{T}$. To derive an explicit form for the coin operator we consider the beam splitter's transformation matrix (see Eq. D.18) given by

$$M_B = \begin{pmatrix} t & r \\ -r^* & t \end{pmatrix}, \tag{3.37}$$

where $t$ and $r$ are the *amplitude* transmission and reflection coefficients respectively. The transformation due to $M_B$ corresponding to the action of the $U_1$ optical unit. Similarly we can derive the transformation matrix for the $U_2$ optical units which utilizes a phase shifter $P_s(\omega_1)$ at the sideward input field and a phase shifter $P_d(\omega_2)$ at the downward output field (see Eq. D.3). Hence the complete transformation due to the cascade of the three optical units is described by

$$\begin{pmatrix} \mathcal{A}_s \\ \mathcal{B}_d \end{pmatrix}_{\text{out}} = \begin{pmatrix} 1 & 0 \\ 0 & e^{-i\omega_2} \end{pmatrix} M_B \begin{pmatrix} e^{-i\omega_1} & 0 \\ 0 & 1 \end{pmatrix} \begin{pmatrix} \mathcal{A}_s \\ \mathcal{B}_d \end{pmatrix}_{\text{in}}, \tag{3.38}$$

where the collective operation of the beam splitter and phase shifters produces the coin matrix

$$\hat{C} = \begin{pmatrix} t\,e^{-i\omega_1} & r \\ -r^*\,e^{-i(\omega_1+\omega_2)} & t\,e^{-i\omega_2} \end{pmatrix}. \tag{3.39}$$

Setting $t = r = 1/\sqrt{2}$, corresponding to a 50:50 beam splitter, as well as $\omega_1 = 0$ and $\omega_2 = \pi$, gives the common Hadamard transformation coin matrix. Similar to the proposal of Zhao et al. (2002), propagating a single photon through the network of optical units produces a final state with probability amplitudes corresponding to the quantum walk distribution. These are then measured by positioning detectors behind the optical units on the last dynamic line as illustrated in Fig. 3.17.

At first this approach would seem unique to a single photon input, since there are only two possible basis states, $|1, 0\rangle$ and $|0, 1\rangle$, which are then intuitively associated with the two coin states $|s\rangle$ and $|d\rangle$. Next however, Jeong et al. (2004) turned their attention to a classical input field and showed that remarkably, the quantum walk can be identically performed without a single photon source – instead using an ordinary laser beam or any other arbitrary fields as the input.

To see this we first consider the quantum description of a radiation field (see Appendix A.2). In the implementation of Jeong et al. (2004), polarization modes are not considered and the modes of the single frequency radiation propagating the network are only spatial. More precisely each row and column in the network has a single spatial mode $\mathbf{k}$ and hence the total number of modes is $2\mathcal{N}$, where $\mathcal{N}$ is the number of nodes on the dynamic line. For a Fock input state the complete state of the system is described by $|n_{\mathbf{k}_1}\rangle|n_{\mathbf{k}_2}\rangle \ldots |n_{\mathbf{k}_{2\mathcal{N}}}\rangle$, where $n_{\mathbf{k}}$ is the number of photons in the spatial mode $\mathbf{k}$. Now when two arbitrary modes, say $\mathbf{k}_2$ and $\mathbf{k}_m$, impinge on a beam splitter, the standard technique for evaluating the output states (see Appendix D.3) is to describe the input as the action of creation operators $\hat{a}_{\mathbf{k}_2}^\dagger$ and $\hat{a}_{\mathbf{k}_m}^\dagger$ on the vacuum state $|0_{\mathbf{k}_1}\rangle|0_{\mathbf{k}_2}\rangle \ldots |0_{\mathbf{k}_m}\rangle \ldots |0_{\mathbf{k}_{2\mathcal{N}}}\rangle$, evolve the creation operators $\hat{a}_{\mathbf{k}_2}^\dagger \longrightarrow \hat{b}_{\mathbf{k}_2}^\dagger, \hat{a}_{\mathbf{k}_m}^\dagger \longrightarrow \hat{b}_{\mathbf{k}_m}^\dagger$ according to the beam splitter transformation matrix $M_B$, and then reconstruct the state using $\hat{b}_{\mathbf{k}_2}^\dagger$ and $\hat{b}_{\mathbf{k}_m}^\dagger$. Figure 3.18 shows this for a number of simple Fock states particularly those with a single photon in the system.

We now contrast this with the case where the radiation from the input source is in a coherent state, closely describing a classical single mode laser beam (see Appendix A.4). It can be shown, using a similar approach to evaluate the output state (see Appendix D.3), that for two coherent states $|\alpha_s\rangle$ and $|\beta_d\rangle$ incident on a beam splitter

$$\begin{pmatrix} \alpha' \\ \beta' \end{pmatrix} = M_B \begin{pmatrix} \alpha \\ \beta \end{pmatrix}, \tag{3.40}$$

corresponding to the optical unit $U_1$. This is depicted in Fig. 3.19. Further, considering the action of a phase shifter on the coherent state (see Eq. D.3) given by

$$\hat{P}(\omega)|\alpha\rangle = |e^{-i\omega}\alpha\rangle, \tag{3.41}$$

leads to the conclusion that for the pair of coherent states incident on the optical unit $U_2$, $\alpha$ and $\beta$ are transformed according to

$$\begin{pmatrix} \alpha' \\ \beta' \end{pmatrix} = \hat{C} \begin{pmatrix} \alpha \\ \beta \end{pmatrix}, \tag{3.42}$$

where the transformation matrix $\hat{C}$ is the same as the coin matrix in Eq. 3.39.

As we saw earlier, in the single photon regime, one can deduce the probability distribution $P(k)$ for the quantum walk by using a line of photon detectors at

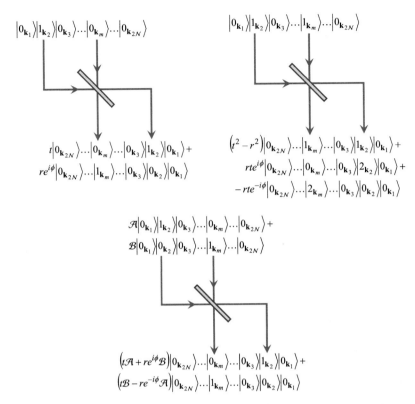

**Fig. 3.18** The action of a beam splitter on simple Fock input states. Parameters $t$ and $r$ are the beams splitter's amplitude transmission and reflection coefficients and $\phi$ represents the phase difference between the reflected and transmitted components

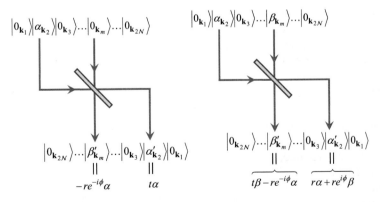

**Fig. 3.19** The action of a beam splitter on coherent input states

the output of the last dynamic line. How does then one measure the probability distribution of the quantum walk in the coherent state regime? Recalling that for a coherent state $|\alpha\rangle$, $|\alpha|^2 = \langle n \rangle$ is the average photon number in that state, the quantum walk distribution can be conveniently obtained through a determination of the output photon count for each node which is proportional to the measured laser intensity. In other words taking $\alpha_0$ to be the single mode coherent input to the network (i.e. all other modes are vacuum states), then the probability distribution $P_k(n)$ at the $k$th node on the dynamic line $n$ is given by

$$
\begin{aligned}
P_k(n) &= \frac{|\alpha_k|^2 + |\beta_k|^2}{|\alpha_0|^2} \\
&= \frac{|E_{k,s}|^2 + |E_{k,d}|^2}{|E_0|^2} \\
&= \frac{I_{k,s} + I_{k,d}}{I_0},
\end{aligned}
\tag{3.43}
$$

where $\alpha_k$ and $\beta_k$ are the coherent output at the $k$th node with corresponding classical electric fields $E_{k,s}$ and $E_{k,d}$, and intensities $I_{k,s}$ and $I_{k,d}$ satisfying $I_0 = \sum_k (I_{k,s} + I_{k,d})$. This corresponds precisely to the distribution expected from a coined quantum walk, simply expressed as normalized laser intensities.

This approach was formally expressed by Mosley et al. (2006); whenever the outcomes are given simply by the outputs of single detectors, they depend only on the second order correlation functions of the inputs, i.e. an intensity-intensity correlation function, providing information on both photon statistics and dynamics of the light generation process of a light source (Choi et al. 2005). Therefore, the results would not differ was the experiment to be performed using a coherent source or repeated using a true single photon source and photon counting detectors.

This is a remarkable result as it clearly demonstrates implementing the quantum walk in a purely classical regime, using classical input (coherent laser), classical linear optics and most importantly a classical measurement scheme. Jeong et al. (2004) went even further to formally show that since any field can be represented as a superposition of coherent states (Glauber 1963) their scheme is indeed universal and is able to take the wave nature of any input field (classical or nonclassical) to show the same interference pattern.

Additionally, Jeong et al. (2004) noted that the required number of resources (in terms of the number of optical elements as well as the field modes involved) grows quadratically with the number of steps. This imposes serious limitations on the scalability of such a proposal and affects the efficiency of simulations using such an interferometric network. The authors then proposed an alternative design represented in Fig. 3.20, in which Acousto-optic modulators (AOM) are used to guide a beam toward a mirror for further steps or toward a detector for measurement. The number of required resources in this latter scheme increases only linearly with the number of steps.

Subsequently Do et al. (2005) constructed an experimental implementation of an optical quincunx based on the design depicted in Fig. 3.21. Despite using PBS and HWP's, similar to the proposal of Zhao et al. (2002), this implementation is considerably simplified and does not require features such as the $\overline{\text{PBS}}$ units and the remixing of $|H\rangle$ and $|V\rangle$ components prior to the rotation by HWP's. In the

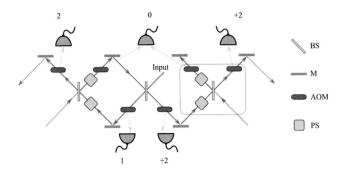

**Fig. 3.20** Alternative setup for quantum walk on a line. In this scheme, the number of required resources scales linearly with the number of steps. Two rows of acousto-optic modulators (AOMs) direct the incoming beams of light to the perfect mirrors M or to the detectors row. This setup is conceptually equivalent to the one sketched in Fig. 3.17b (Adapted from Jeong et al. (2004))

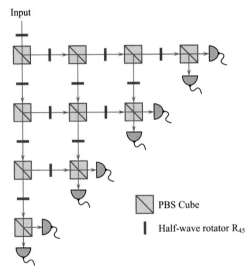

**Fig. 3.21** Idealized implementation of a quantum quincunx by use of optical elements (Adapted from Do et al. (2005))

single photon regime, the coin states are defined using the "sideward" $|s\rangle$ and "downward" $|d\rangle$ modes of the field, introduced by Jeong et al. (2004), and the conditional transition operator $\hat{T}$ is identical to that of Eq. 3.35. Before entering the PBS, the sideward and downward modes each pass through a HWP which rotates the polarization axis. The state of the input laser beams incident on the PBS, before passing through the HWP's, can be written as

$$|\psi_0\rangle = \mathcal{A}|s\rangle + \mathcal{B}|d\rangle$$
$$= \mathcal{A}|s\rangle \otimes |p_s\rangle + \mathcal{B}|d\rangle \otimes |p_d\rangle$$
$$= \mathcal{A}|s\rangle \otimes (h_s|H\rangle + v_s|V\rangle) + \mathcal{B}|d\rangle \otimes (h_d|H\rangle + v_d|V\rangle), \qquad (3.44)$$

where $|p_s\rangle$ and $|p_d\rangle$ are the polarization states of the sideward and downward fields
with $|h_s|^2+|v_s|^2 = |h_d|^2+|v_d|^2 = 1$. Following the unitary rotation $\hat{R}_\theta$ performed
by each of the two HWP's, the new state of the radiation is given by

$$|\psi_{WP}\rangle = \mathcal{A}|s\rangle \otimes \hat{R}_{\theta_s}(h_s|H\rangle + v_s|V\rangle) + \mathcal{B}|d\rangle \otimes \hat{R}_{\theta_d}(h_d|H\rangle + v_d|V\rangle)$$
$$= \mathcal{A}|s\rangle \otimes (h_s'|H\rangle + v_s'|V\rangle) + \mathcal{B}|d\rangle \otimes (h_d'|H\rangle + v_d'|V\rangle). \quad (3.45)$$

The action of the PBS causes the transmission of the $|H\rangle$ components of polariza-
tion and the reflection of the $|V\rangle$ components. Hence

$$|\psi_{PBS}\rangle = |s\rangle \otimes (\mathcal{A}h_s'|H\rangle + \mathcal{B}v_d'|V\rangle) + |d\rangle \otimes (\mathcal{B}h_d'|H\rangle + \mathcal{A}v_s'|V\rangle)$$
$$= \mathcal{A}'|s\rangle \otimes |p_s'\rangle + \mathcal{B}'|d\rangle \otimes |p_d'\rangle$$
$$= \mathcal{A}'|s\rangle + \mathcal{B}'|d\rangle, \quad (3.46)$$

where $|p_s'\rangle$ and $|p_d'\rangle$ are the new polarization states of the sideward and downward
fields. Therefore we may relate the input and output fields according to

$$\begin{pmatrix} \mathcal{A}' \\ \mathcal{B}' \end{pmatrix} = \hat{C} \begin{pmatrix} \mathcal{A} \\ \mathcal{B} \end{pmatrix}, \quad (3.47)$$

where

$$\hat{C} = \begin{pmatrix} t_s & r_d \\ r_s & t_d \end{pmatrix}, \quad (3.48)$$

is in general non-unitary, and $t$ and $r$ are the complex transmission and reflection
coefficients of each field mode which in turn, are related to the coefficients $h$
and $v$ and the transformation $\hat{R}_{2\theta}$ performed by the HWP's (where, as before $\theta$
is the angle between the fast axis of the HWP and the polarization plane). This
lack of guaranteed unitarity exhibited by the operator $\hat{C}$ is indeed an expected
outcome, since the quantum state of light in the network would in fact be naturally
spanned by four basis states $|s, H\rangle$, $|d, H\rangle$, $|s, V\rangle$ and $|d, V\rangle$ for which there is a
unitary $4 \times 4$ transformation relating the input and output superposition states. The
implementation of Do et al. (2005) on the other hand is only concerned with the
projections onto states $|s\rangle$ and $|d\rangle$ for which the $2 \times 2$ transformation $\hat{C}$ may not
be unitary. For example, taking $|\psi_0\rangle = 1/2|s\rangle + 1/2|d\rangle$, if after interacting with
the HWP's the $|s\rangle$ mode entering the PBS is entirely in the $|H\rangle$ polarization state
(i.e. $h_s' = 1$ and $v_s' = 0$) and the $|d\rangle$ mode is entirely in the $|V\rangle$ polarization
state (i.e. $h_d' = 0$ and $v_d' = 1$), then the PBS output will be only $|s\rangle$ with
no downward component, and in this case $\hat{C}$ is clearly non-unitary. On the other
hand by setting $\theta$ to be identical for all HWP throughout the network (which was
the case in the implementation of Do et al. (2005)) so that $h_s = h_d = h$ and
$v_s = v_d = v$, and applying appropriate phase shifts at the PBS, a unitary $\hat{C}$ may
be constructed. Using HWP's with axes oriented at $\theta = 22.5°$ for example leads
to $|h| = |v| = 1/\sqrt{2}$, allowing the PBS to behave like an ordinary 50:50 beam

**Fig. 3.22** Photograph of the
apparatus used by Do et al.
(2005)

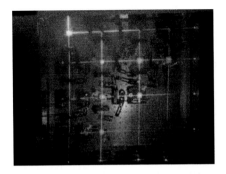

splitter. Including appropriate phase shifts at the PBS, $\hat{C}$ will then be equivalent to
the unitary Hadamard transformation.

Another notable difference between Do et al.'s (2005) implementation and the
original proposal of Zhao et al. (2002) is the use of a low intensity (0.5–1.0 mW)
He-Ne laser (see Fig. 3.22) rather than a genuine single photon source. Nonetheless,
earlier insights provided by Jeong et al. (2004) confirm that the experiment is indeed
valid and that the quantum walker's probability distribution simply corresponds to
the normalized output intensity profile of the network.

More recently, Broome et al. (2010) provided another experimental demonstration of the principle ideas behind the proposal of Zhao et al. (2002), but
with a significant improvement: the number of optical elements required by this
interferometric scheme scales only linearly as $2N$ with the number of steps $N$ (in
contrast to $(N^2 + N)/2$ in Jeong et al. (2004) and Do et al. (2005)). As in that
seminal proposal, quantum coin basis states are encoded in the polarization $|H\rangle$
and $|V\rangle$ of the input photon while the position states are represented by longitudinal
spatial modes marked as $0, \pm 1, \pm 2, \ldots$ in Fig. 3.23.

The coin state of the walk is initialised using the combination of a polarizing
beam splitter as well as quarter and half-wave plates which can be configured to
generate arbitrary polarization states. As usual a half-wave plate at $22.5°$ to the
horizontal is employed to perform the Hadamard coin operation. The innovation
introduced by Broome et al. (2010) is to perform the conditional translations using
birefringent calcite beam displacers (BD) instead of the network of PBS in Zhao
et al.'s original work (2002). The optical axis of each calcite prism is cut so that
vertically polarized light is transmitted directly while horizontal light undergoes
a lateral displacement into a neighboring mode. Considering an input photon in
spatial mode $|i\rangle$, lateral displacement and direct transmission through the BD can
be reinterpreted in the quantum walk's position space as stepping to the left and
right respectively, i.e.

$$|H,i\rangle \xrightarrow{\text{BD}} |V,i-1\rangle \tag{3.49}$$

$$|V,i\rangle \xrightarrow{\text{BD}} |V,i+1\rangle, \tag{3.50}$$

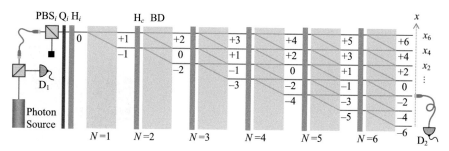

**Fig. 3.23** Experimental schematic for implementing the one-dimensional quantum walk up to the sixth step. A single photon is initially injected into spatial mode 0. Arbitrary initial coin (polarization) states are prepared by the combination of the polarizing beam splitter $PBS_i$ and the quarter and half-wave plates $Q_i$ and $H_i$. Each successive step of the walk involves passage through a half-wave plate $H_c$ performing the coin operation, followed by a calcite beam displacer BD performing the conditional translation. Output photons are coupled to a single mode optic filbert and detected by a single photon detector $D_2$. Coincident detection of photons at detectors $D_2$ and $D_1$ (4.4 ns time window) heralds a successful run of the walk. Constructing the full probability distribution involves repeated detections while incrementally sweeping the $x$ measurement axis (Adapted from Broome et al. (2010))

producing the typical pattern associated with discrete-time walks on a line, where only odd numbered position states are occupied at odd time steps and even sites at even times. At the conclusion of the walk, the probability of finding a photon in position state $i$ is measured by coupling the output photon into an optical fiber, using a fiber coupler placed at location $x_i$ on the measurement axis (see Fig. 3.23), and guided to a single photon detector. The single mode fiber ensures that only one spatial mode is detected at a time. The overall probability distributions shown in Fig. 3.24 is constructed sequentially by repeated photon detections while moving the fiber coupler along the measurement axis in incremental steps.

Pairs of single photons used for this experiment (one to perform the walk and the other for coincident detection) were created via type-II spontaneous parametric down-conversion, at an average rate of $20,000\,s^{-1}$. Hence the mean longitudinal distance between two photons was about 250,000 times longer than the setup length of 60 cm. This together with the $9 \times 10^{-5}$ probability of creating more than one simultaneous photon pair, ensured that only one photon would be in the setup at any given time. The interferometric network itself is inherently stable, where each interferometer can be traced by the beams connecting the $N$th step input in mode $i$ with the $N + 1$th step output also in mode $i$. The transversal modematch is fulfilled because two beams emerging from one displacer will always be parallel, independent of small deviations in the optical alignment. The stability and network alignment procedure are facilitated by the fact that the $N$ interferometers between steps $N$ and $N + 1$ are formed between only two optical components which removes the need for active phase locking.

Another favorable feature arising from the scheme of Broome et al. (2010) is the ability to introduced a tunable decoherence by intentional misalignment of the quantum walk steps as illustrated in Fig. 3.25. More specifically, setting a nonzero

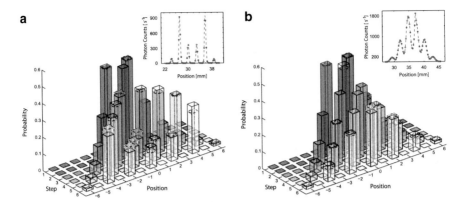

**Fig. 3.24** Probability distributions for successive steps of the (**a**) quantum and (**b**) fully decohered (classical) walks up to the sixth step. *Dashed lines* show experimental data and *solid lines* show theoretical predictions. Probabilities are obtained by normalizing photon counts at each position to the total number of counts for the respective step. The *insets* show horizontal scans across the walk lattice for the five-step quantum walk (coupled into single-mode fiber) and decohered random walk (multimode fiber), respectively (From Broome et al. (2010))

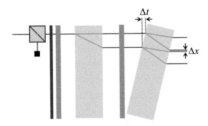

**Fig. 3.25** A relative angle between two beam displacers reduces the recombined photon's temporal ($\Delta t$) and spatial ($\Delta x$) mode overlap, thereby implementing tunable decoherence (Adapted from Broome et al. (2010))

relative angle between neighboring beam displacers leads to both a temporal delay $\Delta t$ and a transversal mode mismatch $\Delta x$ between interfering wave-packets at every step. The experimental setup ensures that neither of these quantities are within the measurement resolution; the coincidence time window is much longer than the temporal shift and the single mode fiber used in photo detection is replaced by a multimode fiber. Hence only the net effect of all the individual misalignments (trace over the temporal and spatial information) appears int the final measurement which corresponds to pure dephasing (Sect. 1.4). In the experiment of Broome et al. (2010), full decoherence occurred at a relative angle of 10.5° at which point the classical random walk probability distribution was recovered as depicted in Fig. 3.24b. The authors also utilised their setup to investigate another qualitative difference between classical and quantum walks, by incorporating absorbing boundaries, implemented using beam blocks at every $i = -1$ spatial mode.

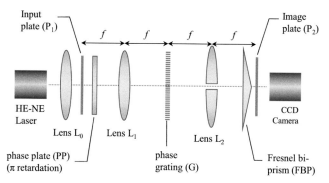

**Fig. 3.26** Experimental setup to obtain one iteration of the quantum walk algorithm (From Francisco et al. (2006b))

### *3.1.5  Optical Refraction*

Francisco et al. (2006b) experimentally demonstrated another all optical implementation of the quantum walk on a line by defining a mapping between the walk states at a certain time and the states of the electromagnetic field incident on a plane. In quantum computation this is the so-called spatial encoding where, different positions in the plane are associated with different computational states and the probability amplitude of the computational state is associated with the (complex) amplitude of the electromagnetic field at the corresponding position. The authors noted that this mapping is exponentially inefficient since in order to represent a quantum state with $n$ qubits it is necessary to divide the plane into $2^n$ regions. However they pointed out that optical simulations of quantum algorithms constitute a simple way to study their basic properties.

In the scheme proposed by Francisco et al. (2006b), each iteration of the walk is performed using the setup illustrated in Fig. 3.26. A 633 nm He-Ne laser beam is expanded, filtered, and collimated by lens $L_0$ so it is possible to choose an almost homogeneous portion of the wave front. The collimated beam impinges onto plane $P_1$ which encodes the state of the walk at step $N$ using a mask generated with a medium where both amplitude and phase of the electromagnetic field can be controlled. Various optical elements then transform the field such that the image produced on plate $P_2$ captured by a CCD camera represents the state of the walk at step $N + 1$. Here the spatial distribution of the beam intensity is associated with the probability to detect each of the states.

Limiting the input plane $P_1$ to a square region, which is split in two halves, the left and right regions correspond to the coin states; left is associated with the state $|0\rangle$ and right is associated with the state $|1\rangle$. Within the left and right regions different slices are defined and labeled by integer numbers $k$. These slices correspond to the position states of the quantum walk labeled by $|0\rangle \otimes |k\rangle$ and $|1\rangle \otimes |k\rangle$ as depicted in Fig. 3.27. Hence the state corresponding to a well defined value for the quantum coin and a well-defined location of the walker is one where the electromagnetic field amplitude is zero everywhere in the plane except for a single slice.

**Fig. 3.27** Representation of the states $|0\rangle \otimes |-1\rangle$, $|0\rangle \otimes |0\rangle$, $|0\rangle \otimes |1\rangle$ (**a**) and $|1\rangle \otimes |-1\rangle$, $|1\rangle \otimes |0\rangle$, $|1\rangle \otimes |1\rangle$ (**b**) of the computational basis as an input scene for the optical modulus (Adapted from Francisco et al. (2006b))

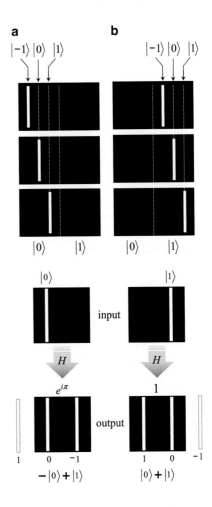

**Fig. 3.28** Schematic demonstration of the Hadamard transformation. Numbers on the white strips in the output states are associated with the diffracted orders. The orders that do not appear in the square image are not registered (*dashed line*) (Adapted from Francisco et al. (2006b))

The coin operator $\hat{C}$ in this scheme is limited to a Hadamard transformation which the authors implemented using a phase plate (PP), a convergent lens $L_1$ used to obtain the optical Fourier transform of the input, a phase grating $G$ and a second lens $L_2$ to produce the inverse Fourier transform. The implementation of the optical Hadamard gate is more thoroughly explored in Francisco et al. (2006a). Figure 3.28 shows the input and output associated with this optical implementation of the Hadamard transformation. When the input intensity corresponds to a slice at a position on the right (representing the state $|1\rangle$), the output intensity is equally distributed between right and left. Similarly, if the input intensity corresponds to a slice on the left (representing the state $|0\rangle$), a superposition of left and right states (with appropriate phases) is generated. The conditional operator $\hat{T}$ is realized by using two shifted halves of the split lens $L_2$. The left half of the lens is displaced to the left and the right half is displaced to the right. By controlling the separation between the two halves it is possible to obtain, in the final plane, a displacement

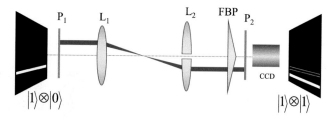

$$|1\rangle \otimes |0\rangle \qquad\qquad\qquad\qquad\qquad\qquad |1\rangle \otimes |1\rangle$$

**Fig. 3.29** Schematic demonstration of the conditional displacement transformation. The *dotted line* on the output plate marks the initial state without the conditional displacement operator (Adapted from Francisco et al. (2006b))

corresponding to one step of the walk (i.e., to move the amplitude from one slice to the next). However, the displacement of the two halves produces an additional linear phase in the final image that is compensated by means of the Fresnel bi-prism (FBP). This process is illustrated in Fig. 3.29 where an upside view of the rays coming from the state $|1\rangle \otimes |0\rangle$ is shown. The rays emerge from the input plane $P_1$, parallel to the optical axis, and which are then focuses by lens $L_1$. In the image plane $P_2$ the beam arrives parallel to the optical axis and shifted by an amount corresponding to the next position in the walk space. The authors noted that this image system produces an inversion that must be taken into account when a measurement is performed.

Francisco et al. (2006b) used their setup to implement the iteration from the third to the fourth step of the quantum walk which is the first state where the quantum and classical pictures diverge. To do so, the initial state $|0\rangle \otimes (|0\rangle + i|1\rangle)/\sqrt{2}$ was numerically iterated three times to compute the state of the walk after the third step. This state was then encoded on a programmable spatial light modulator (with a $640 \times 480$ pixels liquid-crystal display) to create a mask for the input plate $P_1$. Figure 3.30a shows the output image produced by the setup. The authors noted that this image is in part corrupted by the speckle noise due to the high coherence of the laser light, as well as minor aberrations in the optical elements. An intensity profile, obtained by averaging all the rows of that image is shown in Fig. 3.30b and is contrasted with the theoretical distribution in Fig. 3.30c. Figure 3.31 represents the resulting quantum walk probability distribution obtained by the addition of the $|0\rangle$ and $|1\rangle$ profiles in Fig. 3.30b. Although the final probabilities differ slightly from the theoretical values, a good qualitative agreement is obtained.

### 3.1.6  Optical Multiports

Hillery et al. (2003) and Feldman and Hillery (2004) as well as Košík and Bužek (2005) considered multidimensional quantum walks in the context of graphs. They proposed treating the graph as an interferometer where the vertices are optical elements known as multiports. A $2d$-port is a generalized optical element which performs a unitary transformation on $d$ modes of light entering the element through

**Fig. 3.30** Experimental results corresponding to the optical simulation of the quantum walk algorithm from the third to the fourth step. (**a**) Image of the output intensity captured by the CCD. (**b**) Normalized intensity profiles in arbitrary units. (**c**) Theoretical intensity profiles in normalized units (From Francisco et al. (2006b))

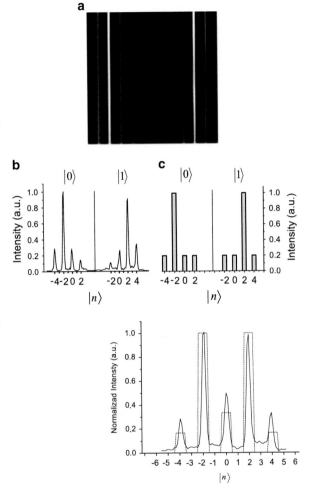

**Fig. 3.31** Final experimental profiles are shown by the *solid line* and the theoretical expected result is shown by the *dashed line* (From Francisco et al. (2006b))

$d$ input ports and exiting through $d$ output ports. Implementing the quantum walk on an arbitrary graph necessitates the use of one $2d$-port corresponding to each vertex, where $d$ is the number of edges meeting at that vertex. Each multiport is specially engineered to comprise $d$ bi-directional ports which simultaneously act as the device input and output ports. In this implementation, graph edges correspond to two way optical pathways for the photons traveling through the interferometer from the output of one multiport to the input of another. Figure 3.32 illustrates a 3D hypercube constructed out of eight 6-ports.

In their original work, Hillery et al. (2003) first introduced a simple implementation of this interferometric quantum walk on a line by using $N$ beam splitters lined up in a row, forming the position states of the walk $|k\rangle$ for $k = 0, 1, 2 \dots N$ from left to right. Consider what happens when a photon traveling in the horizontal direction

**Fig. 3.32** Action of the multiport on the ingoing photon on a 3D hypercube, represented as a unitary operator $\hat{U}_M$. Coherent superposition of three photonic excitations is created. The edges of the hypercube form two-way optical paths (Adapted from Košík and Bužek (2005))

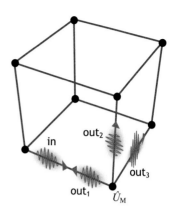

hits a vertical beam splitter. The photon has a certain amplitude to continue in the direction it was going, i.e. to be transmitted, and an amplitude to change its direction, i.e. to be reflected. The beam splitter has two input modes, the photon can enter from either the right or the left, and two output modes, the photon can leave heading either right or left. Hence the internal states of the system can then be described by $|L\rangle$ and $|R\rangle$ corresponding to the photon moving left or right. The beam splitter (see Appendix D.2 for a detailed description) defines a unitary transformation between the input and output modes given by $M_B$ in Eq. D.18 which constitutes the coin operator $\hat{C}$, whereby

$$\hat{C}|k\rangle|R\rangle = t|k\rangle|R\rangle + r|k\rangle|L\rangle, \quad \text{and} \tag{3.51}$$

$$\hat{C}|k\rangle|L\rangle = t^*|k\rangle|L\rangle - r^*|k\rangle|R\rangle. \tag{3.52}$$

If the photon exiting the $k$th node is in state $|L\rangle$, then it naturally moves left to node $k-1$ and if it is in state $|R\rangle$, then it proceeds to node $k+1$. Hence the translation operator $\hat{T}$ is embedded in this implementation, where

$$\hat{T}|k\rangle|R\rangle = |k+1\rangle|R\rangle, \quad \text{and} \tag{3.53}$$

$$\hat{T}|k\rangle|L\rangle = |k-1\rangle|L\rangle. \tag{3.54}$$

Applying the above translation rule to Eq. 3.52 gives

$$\hat{T}\hat{C}|k\rangle|R\rangle = t|k+1\rangle|R\rangle + r|k-1\rangle|L\rangle, \quad \text{and} \tag{3.55}$$

$$\hat{T}\hat{C}|k\rangle|L\rangle = t^*|k-1\rangle|L\rangle - r^*|k+1\rangle|R\rangle. \tag{3.56}$$

Importantly, the authors introduced an alternative perspective of the quantum walk in which the quantum coin is not explicitly present. In this approach, the states are labeled by the edges rather than the vertices in the graph and each edge has two states. If an edge is labeled $j\ k$, with $j$ corresponding to one end and $k$ to the other, then one state is $|j, k\rangle$, corresponding to a photon going from $j$ to $k$, and the other is $|k, j\rangle$, corresponding to a photon going from $k$ to $j$. Now suppose we are in the state $|j - 1, j\rangle$. If the photon is transmitted through the beam splitter it will then be in state $|j, j + 1\rangle$, but if it is reflected it will be in state $|j, j - 1\rangle$. Hence

$$|j - 1, j\rangle \longrightarrow t|j, j + 1\rangle + r|j, j - 1\rangle. \tag{3.57}$$

The other possibility is that the photon is incident on vertex $j$ from the right in which case

$$|j + 1, j\rangle \longrightarrow t^*|j, j - 1\rangle - r^*|j, j + 1\rangle. \tag{3.58}$$

A unitary transformation $\hat{U}$ can be defined such that

$$\hat{U}|j - 1, j\rangle = t|j, j + 1\rangle + r|j, j - 1\rangle, \quad \text{and} \tag{3.59}$$

$$\hat{U}|j + 1, j\rangle = t^*|j, j - 1\rangle - r^*|j, j + 1\rangle, \tag{3.60}$$

which effectively advances the system forward in time a single step. Comparing this with Eq. 3.56, it is easy to see that $\hat{U} \equiv \hat{T} \hat{C}$ and hence the quantum coin is implicit in $\hat{U}$. The authors refer to this new approach as an *edge walk* and noted the analogy with the motion of a particle in a periodic potential where the beam splitters can be thought of as scattering centers with the scattering resulting from a localized potential. They also highlighted a subtle difference between the edge and coined quantum walks. In the coined walk, the probability to be on vertex $j$ is given by combining (taking the squares of the magnitudes and adding) the amplitudes for the states $|j\rangle|R\rangle$ and $|j\rangle|L\rangle$. Under the mapping to the edge walk, these states correspond to states on different edges, $|j - 1, j\rangle$ and $|j + 1, j\rangle$, respectively. However, the probabilities in the edge walk are computed by combining the amplitudes for being on the same edge, e.g., those for $|j - 1, j\rangle$ and $|j, j - 1\rangle$. Therefore, although the overall probability distributions are similar in shape, there will be some differences in the details of the distribution for the two walks.

Hillery et al. (2003) noted that in the interferometer analogy, one can add a new element to the quantum walk that has no analog in classical random walks. Interferometers are made up of multiports and phase shifters; a phase shifter imparts a constant phase $\phi$ to a photon that passes through it (see Appendix D.1). Insertion of a phase shifter into an edge can change the properties of a quantum walk, because it changes how different paths interfere. Suppose that the number of vertices is even, and that we place a phase shifter on all the edges whose left end is an even numbered vertex, i.e. every second edge has a phase shifter on it. The unitary operator $\hat{U}$ that advances this edge walk one step, is now given by

$$\hat{U}|j - 1, j\rangle = te^{i\phi}|j, j + 1\rangle + re^{2i\phi}|j, j - 1\rangle, \quad \text{and} \tag{3.61}$$

$$\hat{U}|j + 1, j\rangle = t^*|j, j - 1\rangle - r^*|j, j + 1\rangle, \tag{3.62}$$

for vertices where $j$ is even, and

**Fig. 3.33** Probability distribution for $N = 50$ and $\phi = p/2$ (From Hillery et al. (2003))

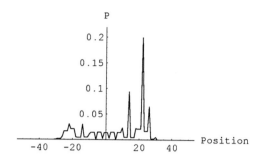

$$\hat{U}|j-1, j\rangle = t|j, j+1\rangle + r|j, j-1\rangle, \quad \text{and} \tag{3.63}$$

$$\hat{U}|j+1, j\rangle = t^* e^{i\phi}|j, j-1\rangle - r^*|j, j+1\rangle, \tag{3.64}$$

for vertices where $j$ is odd. The resulting probability distribution for $\phi = \pi/2$ is depicted in Fig. 3.33. The overall quantum walk signature is clearly present although the detailed characteristics of the distribution are quite different.

The concept of the edge walk provides an intuitive means for extending the quantum walk to graphs of higher dimension where more than two edges emanate from the vertices of the graph. Let the vertex at which all of the edges meet be labeled by $x$ and the opposite ends of the edges be labeled by $y_i$, for $i = 1, 2, 3 \ldots d_x$. In order to construct a walk for a general graph, one chooses a unitary operator $\hat{U}_x$ for each vertex, encapsulating the action of its corresponding multiport which partially reflects and partially transmits an incoming photon (see Fig. 3.32). For any input state $|x, y_i\rangle$ the transition rule dictates that the amplitude for going to the output state $|y_i, x\rangle$ is $r$ and the amplitude for going to any other output state is $t$. That is, the amplitude to be reflected is $r$ and the amplitude to be transmitted through any of the other edges is $t$. Unitarity places two conditions on these amplitudes

$$|r|^2 + (d_x - 1)|t|^2 = 1 \quad \text{and} \tag{3.65}$$

$$(d_x - 2)|t|^2 + r^* t + r t^* = 0, \tag{3.66}$$

and a local unitary operator is then defined such that

$$\hat{U}_x|y_i, x\rangle = r|x, y_i\rangle + t \sum_{\substack{j=1 \\ j \neq i}}^{d_x} |x, y_j\rangle, \tag{3.67}$$

for all $i = 1, 2 \ldots d_x$. The overall unitary operator $\hat{U}$ that advances the walk one step is constructed from the ensemble of local operators $\hat{U}_x$.

Later, Košík and Bužek (2005) used the formalism introduced by Hillery et al. (2003) to implement quantum walks on $d$-cubes, i.e. hypercubes of dimension $d$. Figure 3.34 illustrates multiports arranged in a 3D hypercube with semi infinite tails.

**Fig. 3.34** A 3D hypercube scattering potential. The vertices outside the hypercube are denoted $-\infty, \ldots, -2$ and $d+1, \ldots$ for the hypercube of dimension $d$ (Adapted from Košík and Bužek (2005))

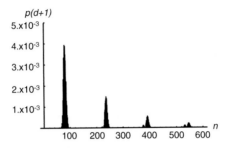

**Fig. 3.35** The scattering probability of a 10-dimensional hypercube, for a photon incoming from the left to be detected by a photon detector placed at the output port of the hypercube (node $d+1$ in Fig. 3.34). $n$ is the number of steps (From Košík and Bužek (2005))

For a photon traveling along one of the tails, the hypercube acts like a scattering potential and hence the authors used the term *scattering quantum walk* to describe this walk. They simulated the probability that an incoming particle from the left will be absorbed by a detector placed along the edge $|d, d+1\rangle$ on the right hand side tail, after some steps. The result for a 10-cube, depicted in Fig. 3.35, shows periodic beating of the probability for the absorption of the photon by the detector.

## 3.2   Nuclear Magnetic Resonance

Du et al. (2003) demonstrated the implementation of a continuous-time walk on a circle with four nodes, using a two-qubit NMR quantum computer. In an NMR quantum computer, each molecule is viewed as a single computer, whose state is determined by the orientations of its spins. Sequences of RF pulses, which manipulate spin orientations and couplings, constitute quantum logic gates and perform unitary transformations on the state. However since direct manipulation of individual spins in a bulk material is not possible, the challenge is to find a global way to use the ensemble of independent quantum computers without needing to address them individually. Gershenfeld and Chuang (1997) showed that this can be done by rearranging (using a series of complex RF pulses) the states of $N$-spin

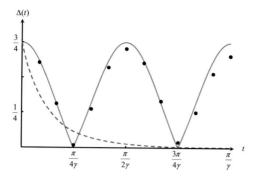

**Fig. 3.36** The total variation distance of the quantum and classical probability distributions from the uniform distribution as a functions of time $t$. The *solid line* corresponds to the quantum distribution $\Delta^Q(t)$ and the *dashed line* to the classical distribution $\Delta^C(t)$, both in theory. The *dots* correspond to the experimental results of the quantum case (From Du et al. (2003))

molecules (which are initially in thermal equilibrium) such that a portion of them forms a uniform background that does not contribute to the measured signal, and the remainder forms a deviation from the mean that will behave as a pure state. Using this technique it is possible to construct logical qubits which are actually collective states of the $N$-spin system.

In the experiment of Du et al. (2003), the quantum computer is implemented using a 0.5 ml, 200 m$M$ sample of carbon-13 labeled chloroform in $d_6$ acetone. The two qubits were defined using the two spin states of $^1H$ and $^{13}C$ nuclei in a magnetic field, with the Hilbert space $\{|\uparrow\rangle \otimes |\uparrow\rangle, |\uparrow\rangle \otimes |\downarrow\rangle, |\downarrow\rangle \otimes |\uparrow\rangle, |\downarrow\rangle \otimes |\downarrow\rangle\}$ representing the four nodes of the walk.

The time-evolution of the continuous-time quantum walk was carried out using the Hamiltonian

$$\hat{H} = \begin{pmatrix} 2\gamma & -\gamma & 0 & -\gamma \\ -\gamma & 2\gamma & -\gamma & 0 \\ 0 & -\gamma & 2\gamma & -\gamma \\ -\gamma & 0 & -\gamma & 2\gamma \end{pmatrix}, \tag{3.68}$$

which is equivalent to

$$\hat{H} = 2\gamma I \otimes I - \gamma(I \otimes \sigma_x + \sigma_x \otimes \sigma_x), \tag{3.69}$$

where $\gamma$ denotes the jumping rate, and $I$ and $\sigma_x$ represent the single qubit's identity and Pauli operators respectively. The continuous-time quantum walk evolution operator $\hat{U} = e^{-iHt}$ was implemented using a sequence of RF pulses, taking $\gamma = \pi J$, where $J = 215$ Hz is the spin-spin couple constant.

The authors showed that in agreement with their theoretical calculations, the evolution of this quantum walk was periodic and reversible, and yielded an exactly uniform probability distribution at certain times (see Fig. 3.36). This is in contrast

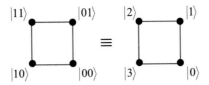

**Fig. 3.37** Logical labeling of the nodes on which the quantum walk is implemented. With this labeling, flipping the first qubit corresponds to a horizontal move and flipping the second qubit a vertical move (Adapted from Ryan et al. (2005))

with the expected classical distribution which is irreversible and only approximates the uniform distribution at infinite-time limit.

Subsequent to the work of Du et al. (2003), Ryan et al. (2005) proposed using a three-qubit liquid-state NMR quantum computer, this time to implement a *coined* quantum walk on a circle with four nodes. For this implementation the authors used the coin operator

$$\hat{C} = \frac{1}{\sqrt{2}} \begin{pmatrix} 1 & 1 \\ 1 & -1 \end{pmatrix}, \tag{3.70}$$

in relation to a coin space with two basis states $|\uparrow\rangle$ and $|\downarrow\rangle$, and defined four nodes labeled $\{|0\rangle, |1\rangle, |2\rangle, |3\rangle\}$ with a conditional translation operator such that

$$\hat{T}|i\rangle|\uparrow\rangle = |i + 1\rangle|\uparrow\rangle, \text{ and} \tag{3.71}$$

$$\hat{T}|i\rangle|\downarrow\rangle = |i - 1\rangle|\downarrow\rangle. \tag{3.72}$$

Experimentally one qubit was used to encode the coin basis states and two qubits for the four position basis states as depicted in Fig. 3.37. The quantum walk was then implemented by formulating the coin and translation operators using *quantum logic gates* – standard ways of manipulating qubits analogous to the way classical logic gates manipulate bits in accordance with the boolean algebra. The coin operator $\hat{C}$ for example can be readily identified as the Hadamard gate. Furthermore using the particular arrangement of qubits in Fig. 3.37 the translation operator can be represented by

$$\hat{T} = (\sigma_x^1 \text{ CNOT}^{1,2} \sigma_x^1) \text{ CNOT}^{1,3}, \tag{3.73}$$

where $\sigma_x$ is the Pauli matrix, CNOT the controlled-NOT gate and superscripts 1, 2 and 3 are the three qubits representing the coin and the two position registers respectively. The complete quantum circuit implementation of the walk is presented in Fig. 3.38. Describing the walk using a series of quantum logic gate operations (i.e. a quantum circuit) is advantageous since the physical implementation of these gates on an NMR quantum computers using a series of FR pulses is well studied (Price et al. 1999).

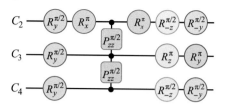

**Fig. 3.38** NMR pulse sequence representing one step of the discrete-time quantum walk. The notation $R_i^\theta$ means a rotation through an angle $\theta$ around the axis $i$ (see Bloch rotations in Appendix C) (Adapted from Ryan et al. (2005))

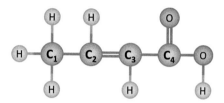

**Fig. 3.39** Molecular structure of trans-crotonic acid (From Ryan et al. (2005))

The experiment was conducted using trans-crotonic acid (four carbons, two hydrogens, and one methyl group) in a solution of deuterated acetone. The $C_3$ carbon atom was used as the coin and $C_2$ and $C_4$ as the position register (Fig. 3.39). $C_1$ was used as an auxiliary to ease the creation of the initial state. Comparing the fidelities of the simulated and experimental results showed good agreement between the two, particularly since the simulations did not account for losses due to state decoherence as well as inhomogeneities in the strong magnetic field and the RF pulses. The authors also demonstrated a convergence of the quantum walk to a classical walk by introducing decoherence of a certain strength between successive steps of the walk.

## 3.3 Cavity QED

Based on the original work of Aharonov et al. (1993), Agarwal and Pathak (2005) described a realization of the discrete-time quantum walk by injecting a single Rydberg atom (i.e. a highly excited atom that resembles hydrogen) into an optical cavity and driving it with a strong continuous external field (Fig. 3.40). The injected atom was then treated as a two-level system, assuming that the ground state $|g\rangle$ and the excited state $|e\rangle$ of the atom were resonant with a single populated cavity mode. In this proposal atomic levels $|g\rangle$ and $|e\rangle$ were taken to be the coin basis states of the walk, while the coherent or Glauber states $|\alpha\rangle$ (see Eq. A.39) of the radiation field inside the cavity formed the position states.

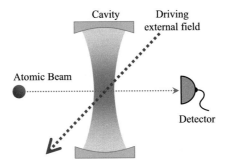

**Fig. 3.40** Schematic diagram of Agarwal and Pathak's quantum walk proposal using an optical cavity. A continuous strong driving field inside the cavity is applied by an external source. The time interval between two atoms in the atomic beam is selected larger than the interaction time in the cavity so that only one atom is present inside the cavity at a time. The atoms are detected at the exit of the cavity by a state-selective detector

Agarwal and Pathak (2005) introduced the Hamiltonian

$$\hat{H} = -i\hbar\chi(\sigma_+\hat{a} - \sigma_-\hat{a}^\dagger) + \hbar(\sigma_+\mathcal{E} + \sigma_-\mathcal{E}^*) \tag{3.74}$$

for the system in the interaction picture (see Appendix B.2.2), where $\chi$ and $\mathcal{E}$ are the coupling constants describing the interaction of the atom with the cavity and the external driving field respectively, $\hat{a}$ and $\hat{a}^\dagger$ are the field's annihilation and creation operators and $\sigma_+ \equiv |e\rangle\langle g|$ and $\sigma_- \equiv |g\rangle\langle e|$ are the atomic raising and lowering operators. Assuming a strong driving field $\mathcal{E}$ and adjusting the phase such that $\mathcal{E}^{*2}/|\mathcal{E}|^2 = 1$, the authors arrived at an effective Hamiltonian

$$\hat{H}_{\text{eff}} = -i\chi\sigma_x(\hat{a} - \hat{a}^\dagger) + 2|\mathcal{E}|\sigma_x, \tag{3.75}$$

where $\sigma_x = (\sigma_+ + \sigma_-)/2$. A closer examination of the effective Hamiltonian reveals that the first term is responsible for producing a displacement in the field state which is dependant on the atomic state, while the second term introduces a rotation in the atomic state. The authors showed that the action of the evolution operator $\hat{U} = \exp(-i\hat{H}_{\text{eff}}t)$ implements exactly the quantum walk as introduced by Aharonov et al. (1993). More precisely, by initializing the atom in the state $c_1|e\rangle + c_2|g\rangle$ and the field inside the cavity in the coherent state $|\alpha\rangle$, we have

$$|\psi(0)\rangle = c_1|e\rangle|\alpha\rangle + c_2|g\rangle|\alpha\rangle, \tag{3.76}$$

and the combined state of the atom-cavity system at time $t$ is given by

$$\begin{aligned}|\psi(t)\rangle &= \hat{U}|\psi(0)\rangle \\ &= c_+e^{-i\phi}(|e\rangle + |g\rangle)|\alpha + \chi t/2\rangle + \\ &\quad c_-e^{i\phi}(|g\rangle - |e\rangle)|\alpha - \chi t/2\rangle, \end{aligned} \tag{3.77}$$

where $c_\pm = c_1 \pm c_2$ and $\phi = (|\mathcal{E}| + \text{Im}(\alpha)\chi/2)\,t$. Hence the action of the evolution operator can be regarded as a sequence of three sub-operators, i.e. $\hat{U} = \hat{R}\hat{T}\hat{C}$ in the following way:

1. The action of $\hat{C}$ on the initial state of the atom-cavity system mixes the amplitudes in the ground and excited states, i.e.

$$|\psi_1\rangle = \hat{C}|\psi(0)\rangle$$
$$= c_+ e^{-i\phi}|e\rangle|\alpha\rangle + c_- e^{i\phi}|g\rangle|\alpha\rangle. \tag{3.78}$$

The matrix

$$\hat{C} = \frac{1}{\sqrt{2}} \begin{pmatrix} e^{-i\phi} & e^{-i\phi} \\ e^{i\phi} & -e^{i\phi} \end{pmatrix} \tag{3.79}$$

is therefore equivalent to the coin operator.

2. The action of $\hat{T}$ on the system produces a conditional displacement in the cavity field state, i.e.

$$|\psi_2\rangle = \hat{T}|\psi_1\rangle$$
$$= c_+ e^{-i\phi}|e\rangle|\alpha + l\rangle + c_- e^{i\phi}|g\rangle|\alpha - l\rangle, \tag{3.80}$$

where $l = \chi t/2$ is the displacement length, assuming $\chi$ is real valued. Operator $\hat{T}$ is therefore equivalent to the conditional translation operator.

3. The action of $\hat{R}$ on the system produces a rotation of the atomic Bloch vector (see Appendix C), i.e.

$$|e\rangle \longrightarrow |g\rangle + |e\rangle \quad \text{and} \tag{3.81}$$
$$|g\rangle \longrightarrow |g\rangle - |e\rangle, \tag{3.82}$$

resulting precisely in the state $|\psi(t)\rangle$ in Eq. 3.77.

The above steps 1 and 2 can be readily identified as the necessary operations for implementing a single iteration of the coined quantum walk. The last step (i.e. the Bloch vector rotation) however becomes important in the context of reading out the final state $|\psi(t)\rangle$ of the system. In this proposal one performs a partial measurement on the final state of the atom-cavity system by detecting whether or not the atom leaving the cavity is in the ground or excited state. However since measuring the superposition state of a quantum system leads to the collapse of its wave-function, it becomes immediately clear that a naive state measurement at the end of step 2 would completely destroy the quantum nature of the walk. In fact as pointed out by Aharonov et al. (1993), without the aforementioned third step, the outcome of this procedure coincides exactly with the distribution expected from a classical random walk. They showed however that by performing a rotation in the atomic Bloch vector it is possible to demonstrate the uniquely quantum characteristics of the walk, while continuing to measure the state of the outgoing atoms. We can see this in Agarwal and Pathak's scheme (2005) by rearranging Eq. 3.77 which yields

$$|\psi(t)\rangle = |g\rangle \left[ c_+ e^{-i\phi} |\alpha + l\rangle + c_- e^{i\phi} |\alpha - l\rangle \right] +$$

$$|e\rangle \left[ c_+ e^{-i\phi} |\alpha + l\rangle - c_- e^{i\phi} |\alpha - l\rangle \right], \qquad (3.83)$$

where $l = \chi t/2$. Thus the detection of the atom in state $|e\rangle$ or $|g\rangle$ leaves the cavity field in a superposition of states $|\alpha \pm \chi t/2\rangle$. Using normalization of atomic states one may choose $c_-/c_+ = \tan(\theta)$ in order to arbitrarily determine the relative likelihood of displacements $l$ and $-l$ in the field state if the atom is detected to be in a particular internal state. Agarwal and Pathak (2005) then showed that if after $N$ steps (i.e. passing $N$ atoms through the cavity, one at a time, without reinitializing the field state and measuring the internal state of the each outgoing atom) all the atoms were detected to be in the atomic state $|g\rangle$, then the state of the radiation field in the cavity is given by

$$|\psi_g(t)\rangle = C \sum_{m=0}^{N} \binom{N}{m} e^{i(N-2m)\phi} \tan(\theta)^{N-m} |\alpha - (N-2m)l\rangle, \qquad (3.84)$$

where $C$ is a normalization factor. The final displacement of the walker after $N$ steps comes as a result of quantum interference among all possible states of the quantum walker. Thus the final displacement depends on the relative weightings $\tan(\theta)^{N-m}$ and the relative phases $\phi$ of these states. This leads to the counter intuitive outcome that even though our translation operation displaces only by $l$, in some cases the particle will jump much further than $l$. In other words, the displacement of the quantum walker is not bounded by the classically possible maximum and minimum displacements $\pm Nl$. Instead, the quantum interference leads to an arbitrary displacement in the quantum walker's position that can be much larger than $\pm Nl$. This is demonstrated in Fig. 3.41 by plotting $\psi_g(x) \equiv \langle x|\psi_g(t)\rangle$ assuming that the initial coherent state $\langle x|\alpha\rangle$ was a simple Gaussian. The displacement of the distribution peak is clearly larger than the width of the initial Gaussian and much larger than $Nl$. A small squeezing in the wave-packet is also generated from the interference effects.

Here it is instructive to note that in their 1993 paper Aharonov et al. (1993) had already proposed a physical scheme for implementing the quantum walk, similar to the work of Agarwal and Pathak (2005). Crucially though, their implementation did not include an external driving source and hence the effective evolution operator for the atom-cavity system did not find an exact mapping to the quantum walk formalism. Nonetheless it did demonstrate, for the first time, the quantum nature of the walk by displaying a larger than classical displacements in the position state of the walk.

In another implementation Sanders and Bartlett (2003) described an alternative approach where, similar to the scheme of Agarwal and Pathak (2005), a two-level

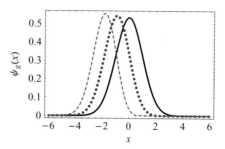

**Fig. 3.41** A plot of $\left|\psi_g(x)\right|$ representing the position of a quantum walker described by Eq. 3.84 after $N = 0, 5$ and 10 steps. The initial distribution is a Gaussian $\exp[-(x - \alpha)^2/2]$ with $\alpha = 0$, the step size $l = 0.05$, $\phi = 2\pi$ and $\theta = 2\pi/3$

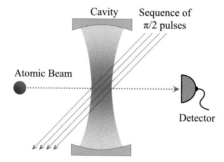

**Fig. 3.42** Schematic diagram of Sanders and Bartlett's quantum walk proposal (2003) using an optical cavity. A single atom traversing through the cavity is subject to periodic Hadamard transformations realized as $\pi/2$ pulses. Between these pulses, the cavity field undergoes a phase shift conditioned on the atomic state

atom traversing a cavity acts as the quantum coin. In contrast however, rather than employing a uniform external driving field, Sanders and Bartlett proposed a periodic sequence of Raman pulses (see Appendix B.3.1), resulting in repeated applications of the coin operator and hence multiple steps of the quantum walk, all during a single passage of the atom through the cavity. This is depicted in Fig. 3.42. Two important assumptions underlying this scheme are

1. The transition time of the atom through the cavity is considerably greater than the Rabi cycle period for the Raman transition, allowing multiple coin operations during one passage, and
2. The action of the transformation itself is instantaneous and does not interfere with the action of the conditional translation operator.

Sanders and Bartlett (2003) adopted the Hadamard coin operator $\hat{C}_H$ implemented as a single $\pi/2$-pulse (see Appendix C.2), although in principle any coin operators may be constructed by carefully tuning the parameters of the Raman transition(s). In their realization, the authors also defined a modified position space for the walk. Whereas previous proposals considered coherent states with real

**Fig. 3.43** Coherent state
Wigner function for
$\alpha = 4.0e^{i\theta}$, where
$\theta = 0, \frac{\pi}{4}, \frac{\pi}{2} \ldots$ Position $x$
is in units of $\sqrt{\hbar/m\omega}$. Up to
$2\pi |\alpha| \approx 25$ packets may be
squeezed on the perimeter of
the circle distinguishable to
within one standard deviation

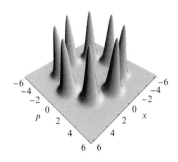

valued $\alpha$ corresponding to a quantum walk in real space, in this scheme complex
coherent states are used to represent the walk in the quadrature phase space. More
specifically, for a given coherent state wave-function $\psi_\alpha(x)$ (see Appendix A.3), the
Wigner function

$$W_\alpha(x, p) = \frac{1}{\pi\hbar} \int_{-\infty}^{+\infty} \psi_\alpha^*(x - y)\psi_\alpha(x + y)e^{2ipy/\hbar}dy \qquad (3.85)$$

produces a localized phase space distribution (with $x$ and $y$ denoting the position
and momentum coordinates) which peaks at $\theta = \arg(\alpha)$ and radius $|\alpha| = \sqrt{\langle n \rangle}$.
This coherent wave-packet has a minimum uncertainty diameter (measured in terms
of quadrature standard deviation) of unity and therefore the circle of radius $|\alpha|$ can fit
approximately $d \lesssim 2\pi |\alpha|$ distinguishable coherent states, as depicted in Fig. 3.43.
The quasi-orthogonal position states of Sanders and Bartlett's quantum walk (2003)
are defined as $|\theta_k = 2\pi k/d\rangle$ for $k = 1, 2 \ldots d$, representing the location of these
non-overlapping packets on the circle. The authors also showed that there is an upper
limit for $|\alpha|$ which effectively places a constraint on the extensibility of the resulting
quantum walk.

To implement the conditional rotation operator, Sanders and Bartlett (2003)
proposed incorporating ac-Stark shifts in the two-level model (Moya-Cessa et al.
1991; Raimond et al. 2001) and requiring that the atomic levels $|g\rangle$ and $|e\rangle$ be highly
detuned from the cavity field. The Hamiltonian for this effect is then given by

$$\hat{H} = \hbar\chi\hat{N} \otimes \hat{\sigma}_z, \qquad (3.86)$$

where $\hat{\sigma}_z$ is the Pauli $z$ matrix, $\chi$ is the atom-cavity coupling strength and the
number operator $\hat{N}$ is the generator of rotation. The Hamiltonian's corresponding
evolution operator is defined by $\hat{R} = \exp(-i\hat{H}\tau/\hbar)$, where $\tau$ is the time between
the application of Hadamard coin operations. The action of the operator $\hat{R}$ on
the combined atom-field system produces the required conditional rotation of the
coherent phase, i.e.

$$\hat{R}|\theta_k\rangle|e\rangle = |\theta_{k-\ell}\rangle|e\rangle, \quad \text{and} \qquad (3.87)$$

$$\hat{R}|\theta_k\rangle|g\rangle = |\theta_{k+\ell}\rangle|g\rangle, \qquad (3.88)$$

**Fig. 3.44** Schematics of a
two-level atom interacting
with the radiation field. The
energy levels $|g\rangle$ and $|e\rangle$ of
the atom are detuned from the
radiation field of frequency $\nu$
by an amount $\delta = \omega_{eg} - \nu$

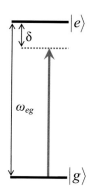

where $\chi\tau = 2\pi\ell/d$. Each step of the quantum walk then evolves according to the
operator $\hat{U} = \hat{R}\hat{C}_H$.

Later, Di et al. (2004) proposed a scheme in which, unlike previous schemes,
the position space of the quantum walk was represented by the Fock states $|n\rangle$
(see Appendix A.3) rather than the coherent states of light. As before, the ground
and excited states of a two level atom ($|g\rangle$ and $|e\rangle$ respectively) provided the coin
basis states. As depicted in Fig. 3.44 however, the frequency $\nu$ of the cavity mode
was detuned from the resonant transition frequency $\omega_{eg}$ by an amount $\delta(t)$. To
implement the quantum walk, the authors made use of the interaction of the two-
level atom with a classical field from an external source, as well as the quantized
field inside the cavity.

In order to implement the coin operator Di et al. (2004) utilized the resonant
Rabi flopping of the two-level atom due to the classical field (see Appendix B.2.1)
for which the evolution operator is given by

$$\hat{U}(\theta, \phi) = \begin{pmatrix} \cos(\theta) & -ie^{i\phi}\sin(\theta) \\ -ie^{-i\phi}\sin(\theta) & \cos(\theta) \end{pmatrix}, \tag{3.89}$$

where $\theta = \Omega\tau$, the interaction energy $\Omega$ is known as the Rabi frequency, $\tau$ is
the interaction time and $\phi$ is the phase of the driving field. The quantum coin was
constructed by tuning the external field parameters such that $\hat{C} = \hat{U}(\frac{\pi}{4}, -\frac{\pi}{2})$.

To describe the conditional translation operator the authors employed the Hamiltonian for the atom-field interaction inside the cavity which, using the dipole approximation (Appendix B.1.3) and the rotating-wave approximation (Appendix B.2.3),
is given by (see Jayne-Cummings model in Appendix B.2.5)

$$\hat{H} = \hbar\nu|e\rangle\langle e| + \hbar\nu\hat{a}^\dagger\hat{a} + \hbar\delta(t)|e\rangle\langle e| + \hbar g_0(|e\rangle\langle g|\hat{a} + \hat{a}^\dagger|g\rangle\langle e|). \tag{3.90}$$

After diagonalizing the Hamiltonian, the atom-field dressed states (energy eigenstates) are given by

$$|+\rangle = \cos(\theta_n)|e\rangle|n\rangle - \sin(\theta_n)|g\rangle|n+1\rangle \quad \text{and} \tag{3.91}$$

$$|-\rangle = \sin(\theta_n)|e\rangle|n\rangle + \cos(\theta_n)|g\rangle|n+1\rangle, \tag{3.92}$$

where

$$\sin(\theta_n) = \frac{\sqrt{\delta^2 + 4g_0^2(n+1)} + \delta}{\sqrt{[\sqrt{\delta^2 + 4g_0^2(n+1)} + \delta]^2 + 4g_0^2(n+1)}} \quad \text{and} \qquad (3.93)$$

$$\cos(\theta_n) = \frac{2g_0\sqrt{n+1}}{\sqrt{[\sqrt{\delta^2 + 4g_0^2(n+1)} + \delta]^2 + 4g_0^2(n+1)}}. \qquad (3.94)$$

It can now be seen that for the atom in state $|+\rangle$ if the detuning is initially, at $t = t_i$, set such that $\delta = -|\delta|$ with $|\delta| \gg 2g_0\sqrt{n+1}$, then $\sin(\theta_n) \approx 0$ and $\cos(\theta_n) \approx 1$ and hence the atom is in state $|e\rangle$ and there are $n$ photons in the cavity. Now if the detuning is chirped slowly such that, at $t = t_f$ (with $|t_i - t_f| \gg 2g_0\sqrt{n+1}$) we have $\delta = +|\delta|$, then $\sin(\theta_n) \approx 1$ and $\cos(\theta_n) \approx 0$ and hence the atom is in the new state $|g\rangle$ and there are $n+1$ photons in the cavity. Using this procedure one can construct an operator $\hat{S}$ such that

$$\hat{S}|e\rangle|n\rangle = -|g\rangle|n+1\rangle \quad \text{and} \qquad (3.95)$$

$$\hat{S}|g\rangle|n\rangle = |e\rangle|n-1\rangle. \qquad (3.96)$$

The conditional translation operator is then given by $\hat{T} = \hat{U}(\frac{\pi}{2}, -\frac{\pi}{2})\hat{S}$, which produces the correct mapping

$$\hat{T}|e\rangle|n\rangle = |e\rangle|n+1\rangle \quad \text{and} \qquad (3.97)$$

$$\hat{T}|g\rangle|n\rangle = |g\rangle|n-1\rangle. \qquad (3.98)$$

In order to control the evolution of the atom due to the external field as well as the time-dependent evolution for chirping during the passage of an atom through the cavity, Di et al. (2004) proposed the atomic levels $|e\rangle$ and $|g\rangle$ to be magnetic sublevels coupled through appropriately polarized light. The interactions can then be controlled via the application of a time-dependent magnetic field such that the interaction times for implementing the $\hat{U}(\theta, \phi)$ transformation and the time dependence of the detuning $\delta(t)$ for the chirping are controlled. The authors also noted that at the time of writing, it was difficult to obtain pure photon number states for large photon numbers.

## 3.4 Quantum Optics

Zou et al. (2006) suggested using photon orbital angular momentum (OAM) to implement the quantum walk on a line (see Appendix A.5). In their proposal the authors employed eigenstates $|l\rangle$ of a single photon in LG modes with $p = 0$ as the

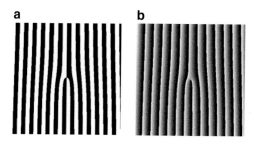

**Fig. 3.45** Computer-generated binary (**a**) and blazed (**b**) templates for computer-generated holograms with a single dislocation. By illuminating these gratings with a Gaussian beam, an $LG_0^1$ mode is produced in the first diffraction order. The diffraction efficiency in the desired order can be increased by blazing (From Vaziri et al. (2002))

position states of the walk. Moving from one position state to another is equivalent to changing the OAM of the photon, which can be achieved using holographic techniques to transform LG modes (Bazhenov et al. 1990; Heckenberg et al. 1992). Conveniently prepared holograms (which are essentially an interference pattern recorded as an optical grating on a transparent film) transform the OAM state of an incoming beam $|l\rangle$ into state $|l+m\rangle$, where $m$ is the number of phase dislocations of the hologram, referred to as the hologram charge. Intuitively, the phase dislocation can be thought of as exerting a 'torque' on the diffracted beam due to the difference in local grating vectors between the upper and lower parts of the grating (Vaziri et al. 2002). As well as the number of dislocations, the 'torque' also depends on the diffraction order. Figure 3.45 shows a computer-generated hologram $Holo_{+1}$ with a single dislocation. When this grating is illuminated with an $LG_0^l$ light, the output mode in the first diffraction order is $LG_0^{l+1}$.

The coin basis states of the walk are represented by two beam paths labeled as the up route $|u\rangle$ and down route $|d\rangle$. The action of the Hadamard coin operator is performed using a 50:50 beam splitter (see Eq. D.18) acting on the single photon input. Figure 3.46 depicts a schematic diagram of the proposed optical network, where each unit $G_i$ performs the $i$th step of the quantum walk. The single photon entering each unit is in a superposition of $|u\rangle$ and $|d\rangle$ states. The beam-splitter performs a unitary (Hadamard) transformation on these coin states. If the photon leaving the beam splitter is in state $|u\rangle$, it is acted upon by $Holo_{+1}$ and if it is in the state $|d\rangle$, it is acted upon by $Holo_{-1}$. This trivially implements the conditional translation operator

$$\hat{T}|l\rangle|u\rangle = |l+1\rangle|u\rangle, \quad \text{and} \tag{3.99}$$

$$\hat{T}|l\rangle|d\rangle = |l-1\rangle|d\rangle. \tag{3.100}$$

Zou et al. (2006) also proposed a similar setup, depicted in Fig. 3.47, for implementing a 1D quantum walk with three internal states $|u\rangle$, $|o\rangle$ and $|d\rangle$, where $|o\rangle$ is an additional third route through which the beam can travel. Here the net effect

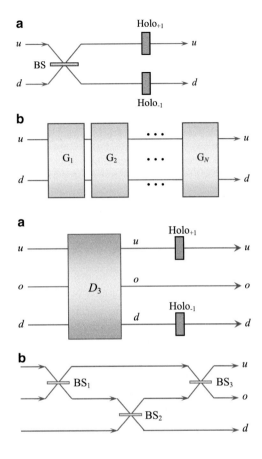

**Fig. 3.46** (a) The implementation of one step of a 1D quantum walk on a line. BS is a symmetric beam splitter. Holo$_l$ denotes the hologram with charge $l = \pm 1$. (b) Optical network with $N$ steps, where G is the experimental setup shown in (a) (Adapted from Zou et al. (2006))

**Fig. 3.47** (a) The implementation of one step of a 1D three-state quantum walk. $D_3$ is a six-port device shown in (b). Holo$_l$ denotes the hologram with charge $l = \pm 1$. (b) The optical implementation of a six-port device $D_3$. The BS$_1$ and BS$_3$ beam splitters have a reflectivity $R = 1/2$; for the other $R = 8/9$ (Adapted from Zou et al. (2006))

of the three beam-splitters BS$_1$ BS$_2$ and BS$_3$ inside each unit D (with reflectivities $\rho_1 = \rho_3 = 1/2$ and $\rho_2 = 8/9$ respectively) is equivalent to the coin operator

$$
\hat{C} = \frac{1}{3} \begin{pmatrix} -1 & 2 & 2 \\ 2 & -1 & 2 \\ 2 & 2 & -1 \end{pmatrix}, \tag{3.101}
$$

and the conditional translation operator obeys

$$
\hat{T}|l\rangle|u\rangle = |l+1\rangle|u\rangle, \quad \text{and} \tag{3.102}
$$

$$
\hat{T}|l\rangle|d\rangle = |l-1\rangle|d\rangle, \quad \text{and} \tag{3.103}
$$

$$
\hat{T}|l\rangle|o\rangle = |l\rangle|o\rangle. \tag{3.104}
$$

The authors point out that their proposed implementations are mainly limited by the diffraction efficiency $p$ of the holograms. Also by increasing the number of

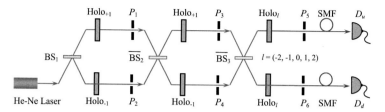

**Fig. 3.48** Experimental setup of an OAM based quantum walk on a line. $BS_1$ is a normal symmetric beam splitter, $\overline{BS}_2$ and $\overline{BS}_3$ are the OAM beam splitters illustrated in Fig. 3.49, $Holo_l$ represents computer generated holograms and $P1$–$P6$ are pinholes used to choose the light beam of proper mode. Results are detected using holograms, single mode fibers (SMF), and power meters $D_u$ and $D_d$ (Adapted from Zhang et al. (2007))

steps $N$, the implementation efficiency decreases exponentially as $p^N$. Therefore cascading the above setup to implement a multi-step quantum walk is an experimental challenge.

Nonetheless Zhang et al. (2007) managed to experimentally demonstrate a three step realization of the quantum walk scheme proposed by Zou et al. (2006). Their experimental setup is shown in Fig. 3.48. As noted earlier (see the implementation of Jeong et al. (2004) in Sect. 3.1), since this experiment depends only on the second order correlation functions of the input (i.e. an intensity-intensity correlation function), it is not necessary to employ a true single photon source and single photon detectors specified by Zou et al. (2006). Hence Zhang and his team used identical coherent states produced by a He-Ne laser of wavelength 632.8 nm. The arm length of the apparatus is about 1.5 m, the period of holograms is 20 lines per mm and pinholes are used to select the first-order diffraction of the holograms.

Zhang et al. (2007) also noted an important subtlety about the use of beam-splitters not discussed in the original proposal of Zou et al. (2006). Since OAM is associated with the helicity of light phase fronts, reflection from a surface will change the OAM states from $l$ to $-l$. Hence, when $LG_0^l$ crosses a symmetric beam splitter, $LG_0^l$ will be exported in the transmitted path and $LG_0^{-l}$ in the reflected path. To realize the quantum walk experimentally, an important modification was then made to the original design, by using a specially devised OAM beam splitter $\overline{BS}$, depicted in Fig. 3.49, instead of a normal symmetric beam splitter BS, to perform the Hadamard transformation.

There have been a number of proposals for measuring the photon OAM (Leach et al. 2002, 2004). In this experiment, to measure the probability of finding the outgoing photons in state $|l\rangle \otimes |u\rangle$ or $|l\rangle \otimes |d\rangle$, a hologram $Holo_{-l}$ was placed in the corresponding route to shift the OAM state to $|0\rangle$ and then the photon was coupled into a single mode fiber (SMF) exploiting the fact that only a photon in the OAM state $|0\rangle$ can be transmitted through an SMF. Subsequently the output energy of the SMF was recorded by a power meter, from which the relevant probabilities could be obtained. These experimental results are listed in Table 3.1. Note that the probabilities of the first two steps are identical to the classical result due to the

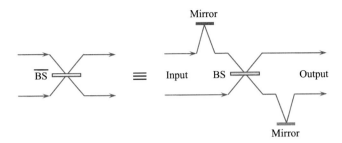

**Fig. 3.49** Implementation of an orbital angular momentum beam splitter $\overline{BS}$. BS is a normal symmetric beam splitter. Mirrors are used here for changing the helical phase of photons, which can translate the OAM state from $l$ to $-l$ (Adapted from Zhang et al. (2007))

**Table 3.1** Experimental probabilities for the first three steps of an OAM based quantum walk (From Zhang et al. (2007))

| Position (3rd step) | −3 | −2 | −1 | 0 | 1 | 2 | 3 |
|---|---|---|---|---|---|---|---|
| Theoretical (%) | 12.5 | 0 | 62.5 | 0 | 12.5 | 0 | 12.5 |
| Experimental (%) | 12.4 ± 0.3 | 0.281 ± 0.005 | 61.4 ± 1.5 | 0.332 ± 0.007 | 13.4 ± 0.3 | 0.136 ± 0.003 | 12.1 ± 0.3 |

absence of interference. The difference of the third step from the classical result however is due to the interference of the $LG_0^0$ mode on $\overline{BS}_3$.

Zhang et al. (2010) later proposed another OAM based quantum walk scheme, making use of the recent invention of an optical device known as the *q-plate*. This allowed the authors to devise a more efficient design without requiring the chain of beam splitters and mirrors, a key source of optical instability in previous schemes (Zou et al. 2006; Zhang et al. 2007) particularly as the number of steps increase. As well as being inherently stable, Zhang et al. (2010) noted that this scheme is also more efficient than previous implementations due to higher transmitted efficiency of q-plates as compared to computer-generated holograms.

Similar to those earlier proposals, this implementation scheme makes use of OAM states $|l\rangle$ to represent the position states of the walk. The coin basis states however are represented by the spin angular momentum (SAM) states denoted by $|L\rangle$ and $|R\rangle$ which correspond to left and right circularly polarized light, carrying $+\hbar$ and $-\hbar$ angular momentum respectively. When light propagates through vacuum or in a homogeneous and isotropic medium, SAM and OAM are independently conserved. In an inhomogeneous and anisotropic medium and under appropriate conditions however, it is possible for light to undergo an exchange of SAM and OAM, in which case only the net angular momentum is conserved. A q-plate is an optical device constructed to perform this function. Physically it comprises a uniaxial birefringent liquid crystal plate whose optical axis is aligned nonhomogeneously in the transverse plane in order to create an effective topological charge $q$ in its orientation. For $q = 1$, incoming photons experience an exchange of $\pm 2\hbar$ angular momentum between SAM and OAM states, meaning that under the action of the q-plate $|L, l\rangle \rightarrow |R, l+2\rangle$ and $|R, l\rangle \rightarrow |L, l-2\rangle$. For $q \neq 1$, there is an additional exchange of $\mp 2\hbar(q-1)$ angular momentum between the plate and OAM.

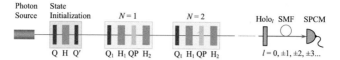

**Fig. 3.50** Experimental scheme for a one dimensional quantum walk using OAM and SAM of single photons. Quarter and half-wave plates Q, Q′ and H can be configured by adjusting the rotation of their optical axis to realize any single bit unitary operation used to initialize the walker's coin state. Each step of the walk is performed using a group of four optical elements: Quarter and half-wave plates $Q_1$ and $H_1$ at $0°$ and $22.5°$ to the horizontal perform the coin operation, while the q-plate QP and another half-wave plate $H_2$ at $0°$ carry out the conditional translation. At the conclusion of the walk, an OAM analysis system, comprising a computer-generated hologram $Holo_l$, a single mode fiber SMF and a single photon counting module SPCM, determines the photon's probability of being in OAM state $l$ (Adapted from Zhang et al. (2010))

Zhang et al.'s (2010) insight was to recognise this property as a convenient building block for constructing a conditional translation operator for the quantum walk.

Figure 3.50 provides a schematic diagram of the proposed implementation scheme. The walk is performed using single photons emitted carrying OAM $l = 0$ and a linear polarization $|H\rangle = (|R\rangle + |L\rangle)/\sqrt{2}$ where $|H\rangle$ denotes the horizontal or $x$ polarization state. Each photon is then passed through an initialization stage comprising a half-wave plate H enclosed by a pair of quarter-wave plates Q and Q′. A variety of initial polarization or coin basis states can be generated by adjusting the orientation of these wave plates' optical axis, as described in Appendix D.4.

Each step of the walk is carried out using a group of four optical elements. A quarter-wave plate $Q_1$ and half-wave plate $H_1$, with their optical axis at $0°$ and $22.5°$ from the horizontal respectively, carry out the quantum coin flip by performing a Hadamard operation

$$|R,l\rangle \xrightarrow{Q_1 \ H_1} \frac{1}{\sqrt{2}}(|R,l\rangle + |L,l\rangle) \qquad (3.105)$$

$$|L,l\rangle \xrightarrow{Q_1 \ H_1} \frac{1}{\sqrt{2}}(|R,l\rangle - |L,l\rangle), \qquad (3.106)$$

on the spin states. This is then followed by the action of a q-plate QP with $q = 0.5$ and an untilted half-wave plate $H_2$

$$|R,l\rangle \xrightarrow{QP} |L,l-1\rangle \xrightarrow{H_2} |R,l-1\rangle \qquad (3.107)$$

$$|L,l\rangle \xrightarrow{QP} |R,l+1\rangle \xrightarrow{H_2} |L,l+1\rangle, \qquad (3.108)$$

performing the conditional translation operator. The quantum walk's final probability distribution is determined similar to earlier implementations, using different computer-generated holograms $Holo_l$ with $l = \pm1, \pm2, \ldots$, a single mode fiber SMF and a single photon count modulator SPCM. The fiber is only transparent to the fundamental mode of light $LG_0^0$ and absorbers all other higher modes. Hence a photon in OAM state $|l\rangle$ can be detected by using a hologram Holo$_{-l}$ which changes its OAM state to $|0\rangle$ before it is passed through the SMF and detected by the SPCM.


## 3.5  Ion Traps

Travaglione and Milburn (2002) proposed the use of an ion trap based on the work of Monroe et al. (1996), where a single $^9$Be$^+$ ion was laser-cooled (Monroe et al. 1995) and confined in a coaxial resonator radio frequency ion trap. Although Monroe and his team had set out to demonstrate the creation of a superposition or "Schrödinger cat" state in a mesoscopic system, the procedure they used was later found to be the same steps necessary to implement a coined quantum walk.

Figure 3.51 depicts the energy levels of the $^9$Be$^+$ ion which has undergone hyperfine splitting due to the presence of a magnetic field. It is these energy levels which provide the necessary mapping to the position and coin states of the quantum walk in the following way:

- There are two internal (electronic) states used to represent the quantum coin states: the stable $^2S_{1/2}(F = 2, m_F = -2)$ and $^2S_{1/2}(F = 1, m_F = -1)$ hyperfine ground states denoted by $| \downarrow\rangle_i$ and $| \uparrow\rangle_i$ respectively, where $F$ and $m_F$ are quantum numbers representing the total internal angular momentum of the atom and its projection along a quantization axis. These states are separated in frequency by $\omega_{HF}/2\pi \approx 1.250\,\text{GHz}$.
- The external (motional) states of the ion which are characterized by the quantized vibrational harmonic oscillator states $|n\rangle_e$, in the $x$-dimension, where $n = 0, 1, 2 \ldots$. These motional states are separated in frequency by $\omega_{HF}/2\pi \approx 11.2\,\text{MHz}$. A natural basis for representing the position states of the walk are the coherent superposition states $|\alpha\rangle$ described by Eq. A.39, where the basis Fock states $|n\rangle$ represent the vibrational electronic states $|n\rangle_e$ under consideration here. Note that coherent states correspond to localized wave-packets moving in real space as depicted in Fig. A.1.

A key feature of this scheme is the use of spin(state)-dependent optical dipole force (Wineland et al. 2003; Schmitz et al. 2009) to split the localized wave-packet. Such a force can be created using a pair of Raman laser beams (see Appendix B.3.1) with the appropriate choice of polarizations (see Appendix E.3 for the origins of a similar force used for state(spin)-dependant transport in neutral atoms). For a coherent wave-packet $|0\rangle$ in the ground state, the resulting spin(state)-dependent displacement $\pm\alpha$ in the position of the wave-packet in real space can the be described as a spin(state)-dependent shift in the phase space of the coherent wave-function (see Fig. A.1b) described using

$$|\alpha\rangle = \hat{D}(\alpha)|0\rangle, \tag{3.109}$$

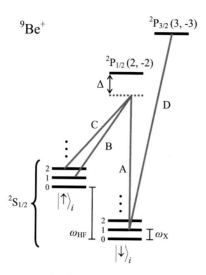

**Fig. 3.51** Electronic structure of the $^9$Be$^+$ ion. Hyperfine ground states $^2S_{1/2}(F = 2, m_f = -2)$ and $^2S_{1/2}(F = 1, m_f = -1)$ are denoted by $|\uparrow\rangle_i$ and $|\downarrow\rangle_i$. Raman beams $A$, $B$ and $C$ are detuned by $\Delta$ from the $^2P_{1/2}(F = 2, m_f = -2)$ excited state which acts as the virtual level providing the Raman coupling. The external motional states are characterized by the quantized vibrational harmonic oscillator states $|n\rangle_e$, in the $x$-dimension. The detection beam $D$ is resonant with the cycling $|\downarrow\rangle_i \longrightarrow {}^2P_{1/2}(F = 3, m_f = -3)$ transition after which the resulting ion fluorescence is observed (Adapted from Monroe et al. (1996))

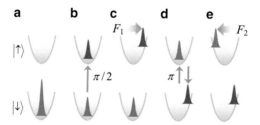

**Fig. 3.52** The evolving state of the $^9$Be$^+$ system following the application of sequential pulses of the Raman beams (Adapted from Monroe et al. (1996))

where $\hat{D} \equiv \exp(\alpha\hat{a}^\dagger - \alpha^*\hat{a})$ is the displacement operator, the displacement $\alpha = \eta\Omega_d\tau$, $\eta$ is the Lamb-Dicke parameter and $\Omega_d/2\pi$ is the coupling strength of the displacement Raman beams.

In the experimental procedure of Monroe et al. (1996), the ion was initially cooled to the motional and electronic ground state $|0\rangle_e|\downarrow\rangle_i$, illustrated in Fig. 3.52a. Then employing a series of stimulated $\Lambda$ Raman transitions with $^2P_{1/2}(F = 2, m_F = -2)$ as the intermediary upper state and detuning $\Delta \approx -12\,\text{GHz}$, Monroe and his team arrived at the superposition state

$$|\alpha\rangle_e|\downarrow\rangle_i + |-\alpha\rangle_e|\uparrow\rangle_i \qquad (3.110)$$

after the following four steps:

1. Applying the $\pi/2$-pulse carrier beams A and B depicted in Fig. 3.51 (with a relative phase difference $\mu$) to drive a Bloch rotation in the internal state of the atom (see Appendix C.2.2), transforming the initial ground state to the superposition state

$$|\psi^1\rangle = \frac{1}{\sqrt{2}} \left( |0\rangle_e| \downarrow\rangle_i - i e^{-\mu} |0\rangle_e| \uparrow\rangle_i \right), \tag{3.111}$$

   as illustrated in Fig. 3.52b.
2. Applying the displacement beams B and C (with a relative phase difference $-\phi/2$) to activate a spin(state)-dependant phase shift, exciting only the motion correlated with the $|\uparrow\rangle_i$ component to a coherent state $|\alpha e^{-i\phi/2}\rangle_e$, hence resulting in the new state

$$|\psi^2\rangle = \frac{1}{\sqrt{2}} \left( |0\rangle_e| \downarrow\rangle_i - i e^{-\mu} |\alpha e^{-i\phi/2}\rangle_e| \uparrow\rangle_i \right), \tag{3.112}$$

   as illustrated in Fig. 3.52c. This selective behavior is achieved by making the C beam $\sigma^-$-polarized which prevents it from coupling the internal state $| \downarrow\rangle_i$ to any virtual $^2P_{1/2}$ states. The selectivity of the displacement force provides quantum entanglement of the internal state with the external motional state.
3. Applying the $\pi$-pulse carrier beams A and B (with a relative phase difference $\nu$) to swap the internal states of the superposition, i.e.

$$|\psi^3\rangle = \frac{1}{\sqrt{2}} \left( e^{i(\nu-\mu)} |\alpha e^{-i\phi/2}\rangle_e| \downarrow\rangle_i + i e^{-i\nu} |0\rangle_e| \uparrow\rangle_i \right), \tag{3.113}$$

   as illustrated in Fig. 3.52d.
4. Applying the displacement beams B and C (with a relative phase difference $\phi/2$) again to activate another spin(state)-dependant phase shift, this time exciting the motion correlated with the $| \uparrow\rangle_i$ component to a coherent state $|\alpha e^{i\phi/2}\rangle_e$, hence

$$|\psi^4\rangle = \frac{1}{\sqrt{2}} \left( e^{i(\nu-\mu)} |\alpha e^{-i\phi/2}\rangle_e| \downarrow\rangle_i + i e^{-i\nu} |\alpha e^{i\phi/2}\rangle_e| \uparrow\rangle_i \right), \tag{3.114}$$

   as illustrated in Fig. 3.52e.

Using the $\sigma^-$-polarized detection beam D and observing the ion fluorescence, the team were able to verify this superposition.

These steps were later identified by Travaglione and Milburn (2002) as forming the procedure for performing a single step of the quantum walk on a line. More precisely in their proposal, they implied that step 1 of the above would correspond to the coin operator

$$\hat{C} = \frac{1}{\sqrt{2}} \begin{pmatrix} 1 & 1 \\ -1 & 1 \end{pmatrix}, \tag{3.115}$$

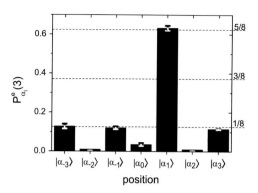

**Fig. 3.53** Experimentally determined distribution of occupation probabilities $P^e_{\alpha_i}(3)$ of the positions in phase space, after step 3 of the quantum walk. Ideally (classically as well as quantum mechanically), we expect the populations $P^e_{\alpha_i}(3) = 0$ for even $i$ and $P^e_{\alpha_{\pm 3}}(3) = 1/8$ of the extremal positions each accessible via 1 out of 8 paths only. Classically we expect $P^e_{\alpha_{\pm 1}}(3) = 3/8$. For an optimized step duration we obtain probabilities very close to the theoretically expected and realistically simulated values of the quantum walk, $P^e_{\alpha_{-1}}(3) = 1/8$ and $P^e_{\alpha_1}(3) = 5/8$ (From Schmitz et al. (2009))

which mixes the coin states of the walk $|\downarrow\rangle_i$ and $|\uparrow\rangle_i$. Subsequently by setting $\phi = \pi$ and $\mu = \nu = \pi/2$, steps 2–4 would implement the conditional translation operator $\hat{T}$ which acts on the position states of the walk $|\alpha\rangle$ to implement the transformation

$$\hat{T}|0\rangle_e|\uparrow\rangle_i = |\alpha\rangle_e|\uparrow\rangle_i \quad \text{and} \tag{3.116}$$

$$\hat{T}|0\rangle_e|\downarrow\rangle_i = |-\alpha\rangle_e|\downarrow\rangle_i. \tag{3.117}$$

Later Xue et al. (2009) pointed out a number of practical drawbacks in the proposal of Travaglione and Milburn (2002) including the readout procedure, and advanced a number of techniques to circumvent them.

Around the same time Schmitz et al. (2009) implemented an experimental proof of principle for a quantum walk in phase space, using a $^{25}$Mg$^+$ ion in a linear multizone Paul trap. Unlike the proposal of Travaglione and Milburn (2002) however, the asymmetric coin given in Eq. 3.115, was implemented using a resonant radio frequency pulse to drive Rabi flopping (see Appendix C.2.1) between the internal levels $|F = 3, M_F = 3\rangle$ and $|F = 2, M_F = 2\rangle$ of the ion. The two displacement Raman laser beams responsible for implementing the conditional translation operator were both blue detuned ($2\pi \times 80$ GHz) from the $^2S_{1/2} \longrightarrow ^2 P_{3/2}$ transition ($\lambda = 280$ nm). Schmitz et al. (2009) described a series of RF and Raman pulses to determine the motional state populations $P^e_{\alpha_i}(N)$ after finishing the $N$th step of the walk. Figure 3.53 demonstrates the resulting distribution after three steps, each performed with a fidelity exceeding 0.99. The authors also proposed various means for increasing the number of steps with comparable fidelity, but noted the challenge posed by the readout of position states for a large number of steps.

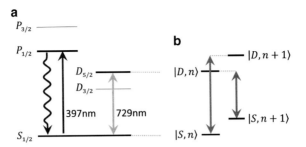

**Fig. 3.54** Electronic structure of a $^{40}$Ca$^+$ ion. (**a**) Internal states $|-\rangle_z \equiv |S_{1/2}, m = 1/2\rangle$ and $|+\rangle_z \equiv |D_{5/2}, m = 3/2\rangle$ used to encode the coin space of the quantum walk. The $S_{1/2} \longleftrightarrow P_{1/2}$ transition is employed for fluorescence state detection. The conditional translation operator utilises a bichromatic light field at 729 nm that is resonant with both the *blue* and *red* axial sidebands (**b**) of the $|-\rangle_z \longleftrightarrow |+\rangle_z$ transition

More recently Zähringer et al. (2010) implemented another experimental demonstration of a quantum walk in which the walker's position is represented by the motional states $|x\rangle$ of a $^{40}$Ca$^+$ ion, while its coin basis states $|\pm\rangle_x$ correspond to projections along the $x$-axis of the ion's internal states $|-\rangle_z \equiv |S_{1/2}, m = 1/2\rangle$ and $|+\rangle_z \equiv |D_{5/2}, m = 3/2\rangle$ projected along the $z$-axis. This is equivalent to a $\pi/2$ rotation of the $|\pm\rangle_z$ components about the $y$-axis of the Bloch sphere (see Appendix C), meaning $|\pm\rangle_x = (|-\rangle_z \pm |+\rangle_z)/\sqrt{2}$.

The conditional translation operator is realised using a bichromatic light field at 729 nm that is resonant with both the blue and red axial sidebands of the $|-\rangle_z \longleftrightarrow |+\rangle_z$ transition as depicted in Fig. 3.54. The resulting Hamiltonian in the Lamb-Dicke regime is given by

$$H_D = \hbar\eta\Omega \left\{ (\sigma_x \cos\phi_+ - \sigma_y \sin\phi_+) \otimes ((\hat{a}^\dagger + \hat{a})\cos\phi_- + i(\hat{a}^\dagger - \hat{a})\sin\phi_-) \right\} \tag{3.118}$$

where $\eta = 0.06$ is the Lamb-Dicke parameter, $\Omega$ is the Rabi frequency, and $2\phi_\pm = \phi_b \pm \phi_r$ with $\phi_b$ and $\phi_r$ being the phases of the light fields tuned to the blue and red sidebands respectively. By adjusting the phase parameters according to $\phi_- = \pi/2$ and $\phi_+ = 0$ the Hamiltonian is reduced to

$$H_d = 2\eta\Omega\Delta_x \sigma_x \hat{p} \tag{3.119}$$

where $\hat{p} = i\hbar(\hat{a}^\dagger - \hat{a})/2\Delta_x$ is the momentum operator with $\Delta_x = \sqrt{\hbar/2m\omega_{xa}}$ and $\omega_{xa} = 2\pi \times 1.356$ MHz being the linear Paul trap's axial frequency. Examining the resulting evolution operator given by

$$\hat{T} = \exp(-\frac{i}{\hbar}\sigma_x \hat{p}\, d), \tag{3.120}$$

it can be recognized as implementing a state-dependent translation operator where $\hat{T}|x, \pm\rangle = |x \pm d, \pm\rangle$ coherently splits the ion's wave-function in phase space along the $x$-axis, where $d = 2\eta\Omega\tau\Delta_x$ is the step size by which the two emerging wave-packets $\psi_\pm$ move in opposite directions. For a $\tau = 40\,\mu s$ pulse with a Rabi frequency of $\Omega = 2\pi \times 68\,kHz$, this step size is long enough to ensure that the resulting wave-packets remain orthogonal $|\langle\psi_+|\psi_-\rangle| \approx 0.02$ while allowing for a large number of steps in phase space.

A symmetric coin operator is implemented using a $\pi/2$ pulse acting on the carrier wave. This corresponds to the action of the operator

$$\hat{C} = \exp(-i\frac{\pi}{4}\sigma_z), \tag{3.121}$$

which creates an equal superposition of $\sigma_x$ eigenstates for both wave-packets $\psi_+$ and $\psi_-$. Zähringer et al. (2010) initialised the quantum walk by preparing the ion in the state $|+\rangle_y = (|+\rangle_z + i|-\rangle_z)/\sqrt{2}$ and proceeded to perform a symmetric walk for up to 13 steps.

The measurement process involves readjusting the phase parameters in Eq. 3.118 according to $\phi_- = \phi_+ = 0$ to produce another state-dependant displacement operation given by

$$U_p = \exp(-i\frac{k}{2}\sigma_x\hat{x}), \tag{3.122}$$

where $\hat{x} = (\hat{a}^\dagger + \hat{a})\Delta_x$ is the position operator and $k = 2\eta\Omega_p t/\Delta_x$ is proportional to the interaction time $t$. The application of the propagator $U_p$ followed by a measurement of $\sigma_z$ is equivalent to measuring the observable

$$O(k) = U_p^\dagger\sigma_z U_p = \cos(k\hat{x})\sigma_z + \sin(k\hat{x})\sigma_y. \tag{3.123}$$

Hence if the ion's internal state is $|+\rangle_y$ the measured quantity corresponds to $\langle O(k)\rangle = \langle\cos(k\hat{x})\rangle$ and if it is $|+\rangle_z$ the measurement gives $\langle O(k)\rangle = \langle\sin(k\hat{x})\rangle$. A Fourier transformation of these measurements yields the probability density $\langle\delta(\hat{x} - x)\rangle$ in position space.

In order to perform the measurement all internal state populations are first recombined in $|-\rangle_z$ by first transferring the population in $|-\rangle_z$ to the auxiliary state $|D_{5/2}, m = 5/2\rangle$, then transferring the population in $|+\rangle_z$ back to $|-\rangle_z$, and finally exciting the population from $|D_{5/2}, m = 5/2\rangle$ to $|P_{3/2}, m = 3/2\rangle$ from where it spontaneously decays to $|-\rangle_z$ with efficiencies >99%. By transferring the total recombined population in $|-\rangle_z$ to the internal state $|+\rangle_z$, Zähringer et al. (2010) were able to producing the even Fourier components $\langle\cos(k\hat{x})\rangle$ by making a series of $\langle\sigma_z\rangle$ measurements corresponding to probe times $t$ between 0 and $300\,\mu s$ and $\Omega_p = 2\pi \times 26\,kHz$ with state detection performed via fluorescence detection on the $S_{1/2} \longleftrightarrow P_{1/2}$ transition (Fig. 3.54). Likewise odd Fourier components $\langle\sin(k\hat{x})\rangle$ were determined after transforming the recombined population in $|-\rangle_z$ to internal state $|+\rangle_y$ using a $\pi/2$ pulse. The measured Fourier components and the resulting probability density $\langle\delta(\hat{x} - x)\rangle$ are displayed in Fig. 3.55 for a seven step walk.

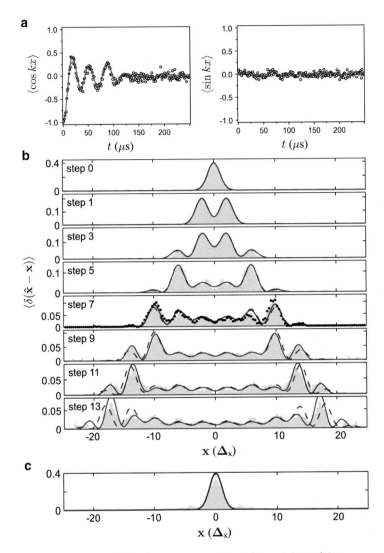

**Fig. 3.55** (**a**) Measurement of Fourier components $\langle\cos(k\hat{x})\rangle$ and $\langle\sin(k\hat{x})\rangle$ for a seven step quantum walk. The data are obtained by varying the duration of the probe pulse for the ion prepared in the internal state $|+\rangle_z$ (*left*) or $|+\rangle_y$ (*right*) after completing the walk. The probability distribution is obtained by Fourier transforming a fit to the data (*solid line*). (**b**) Reconstruction of the symmetric part of the probability distribution $\langle\delta(\hat{x}-x)\rangle$ for up to 13 steps in the quantum walk. The *blue dashed curve* is a numerical calculation for the expected distribution within the Lamb-Dicke regime. The *blue solid curve* takes into account corrections to the Lamb-Dicke regime. In step 7, the *dotted curve* represents the full reconstruction using also the $\langle\sin(k\hat{x})\rangle$ shown in (**a**). (**c**) Probability distribution of a five step quantum walk after the application of five additional steps which reverse the walk and bring it back to the ground state (From Zähringer et al. (2010))

Finally, Zähringer et al. (2010) extended the quantum walk by adding a second ion to the system, thus introducing a four sided coin space in which states $|+, +\rangle_x$ and $|-, -\rangle_x$ correspond to conditional translations to right and left respectively, while states $|+, -\rangle_x$ and $|-, +\rangle_x$ remain stationary. For the two-ion quantum walk all pulses are applied to both ions simultaneously and the probability distribution of the center-of-mass mode is obtained in the same way as for a single ion.

## 3.6  Neutral Atom Traps

Neutral atom traps (Yin 2006) are often produced using a standard protocol in which the atoms are first cooled and collected in a magneto-optical trap (MOT) and then transferred into a conservative, either magnetic or optical trap. A number of authors (Chandrashekar 2006; Eckert et al. 2005; Dur et al. 2002; Joo et al. 2007; Côté et al. 2006; Manouchehri and Wang 2009) have proposed using the latter trapping mechanism, i.e. an all optical trap, for implementing the quantum walk.

A *dipole trap* is formed as a result of the interaction between the atom(s) and the radiation field of a single laser beam. But it is also possible to introduce a second laser field where the resulting interference produces a *dipole lattice trap* or *optical lattice* described in Appendix E. Important differences between dipole and lattice traps are discussed in Appendix E.1.

Irrespective of the confinement type, the atomic cloud can be further cooled to form a Bose-Einstein condensate (BEC); a mesoscopic coherent state of matter in which all atoms show *identical quantum properties*. A BEC can be imaged without being destroyed and its interactions with laser light are much stronger than that of a single particle (Jaksch 2004). Such favorable properties have promoted the use of trapped BEC's in a number of quantum walk realization schemes.

### 3.6.1  Dipole Traps

Chandrashekar (2006) proposed one such implementation using a BEC of $^{87}$Rb atoms transferred to a far detuned optical dipole trap with a long Rayleigh range $\pm Z_R$ from the focal point in the axial direction of the beam. In this proposal the position states of the walk are described by the wave-packet $|\psi_x\rangle$ localized around a position $x$, while the coin basis states of the walk are taken to be the hyperfine states of the BEC in its ground state defined by

$$|0\rangle = |F = 1, m_f = 1\rangle \tag{3.124}$$

$$|1\rangle = |F = 2, m_f = 2\rangle. \tag{3.125}$$

In his proposed scheme Chandrashekar (2006) suggested implementing the conditional translation operator by giving the trapped BEC a *momentum kick* to the left or right depending on its internal state. The state dependent momentum kick

Fig. 3.56 Light field
configuration for a stimulated
Raman transition process to
give directional momentum
kick. $\Delta$ is the detuning of the
laser from its transition
frequency. $|S\rangle$ signifies
$1/\sqrt{2}(|0\rangle \pm |1\rangle)$ (Adapted
from Chandrashekar (2006))

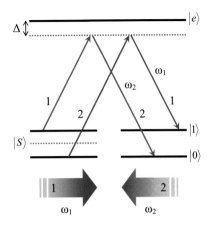

is the product of a stimulated Raman process (see Appendix B.3.1), whereby a pair
of counter propagating laser beams 1 and 2 with frequencies $\omega_1$ and $\omega_2$ and wave
vectors $k_1$ and $k_2$ (along the axial direction of the optical trap) are tuned to drive the
transition between the internal states $|0\rangle$ and $|1\rangle$ via an intermediary virtual state $|e\rangle$
(see Fig. 3.56). This results in a coherent exchange of photons between the two laser
fields involving either the absorption of a photon from laser field 1 and stimulated
emission into laser field 2, or absorption from field 2 and stimulated emission into
field 1. This inelastic stimulated Raman scattering process imparts a well-defined
momentum on the coherent atoms;

$$P = \hbar(k_1 - k_2) = \hbar\delta_z \qquad (3.126)$$

to the left during $|0\rangle \xrightarrow{\omega_1} |e\rangle \xrightarrow{\omega_2} |1\rangle$ and

$$P = \hbar(k_2 - k_1) = -\hbar\delta_z \qquad (3.127)$$

to the right during $|1\rangle \xrightarrow{\omega_2} |e\rangle \xrightarrow{\omega_1} |0\rangle$, where

$$|k_2 - k_1| = |k_1 - k_2| = \delta_z. \qquad (3.128)$$

Thus the BEC receives a momentum kick $\hbar\delta_z$ or $-\hbar\delta_z$, conditioned on being in the
coin state $|0\rangle$ or $|1\rangle$. This process, known as *stimulated Raman kick*, was originally
developed by Hagley et al. (1999) and experimentally used to extract sodium atoms
from a trapped BEC with control on the direction of the outgoing atoms.

The Raman transition proceeds according to the RWA Hamiltonian given by
Eq. B.87. With the appropriate choice of interaction energies $\Omega_1$ and $\Omega_2$, detuning
$\Delta$ and two-photon transition time $t$, it is possible to engineer an evolution operator
$\hat{U}(t) = \exp(i\hat{H}t/\hbar)$ which closely approximates a $\pi$ Bloch rotation $\hat{R}_\pi \equiv$
$|1\rangle\langle2| + |2\rangle\langle1|$, causing the coin states of the atom to be swapped. This effect is
desirable since it can be easily reversed via a compensatory bit flip, but this time
implemented using an RF $\pi$ pulse which achieves a direct Rabi transition between

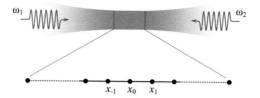

**Fig. 3.57** Positions $x$ of a BEC wave-packet subject to Stimulated Raman kicks, imparted as a result of the interaction between a pair of counter propagating laser beams with frequencies $\omega_1$ and $\omega_2$ (Adapted from Chandrashekar (2006))

states $|0\rangle$ and $|1\rangle$ without imparting a momentum change. Appendix C.2 describes how Rabi flopping and Raman transitions can be tailored to implement generalized Bloch rotations.

Starting with an $N$ atom BEC in state

$$|\psi_{\text{init}}\rangle = \frac{1}{2}\left(|0\rangle + |1\rangle\right) \otimes |\psi_{x_0}\rangle, \tag{3.129}$$

the conditional translation operator $\hat{T}$ is implemented by subjecting the atoms to a pair of counter propagating beams to implement the stimulated Raman kick for a duration $t = P/(ml)$, the time taken by the condensate with mass $m$ to move a distance $l$ (Fig. 3.57). During or after the conditional translation the RF $\pi$ pulse is applied to compensate for the unwanted coin flip. Finally a unitary rotation of the internal states or the coin operator can be realized by applying an RF pulse, fast laser pulse, a standard Raman pulse, or microwave techniques.

It should be noted here that according to Chandrashekar (2006), in order to implement a Hadamard walk, the spatial width of the particle has to be larger than the step length $l$. As already discussed in Sect. 1.1.2 however, the position space of the walk is required to be (quasi-)orthogonal (i.e. the wave-packet width $\lesssim l$) if it is to exhibit the characteristics of contemporary quantum walks.

The number of quantum steps that can be implemented using the BEC without decoherence would depend on two main factors: (a) the Rayleigh range of the dipole trap. As the BEC moves away from the trap center the width of the BEC wave-packet increases and contributes to the internal heating of the atoms. Beyond a certain distance $x_n$ from the trap center, atoms in the trap decohere resulting in the collapse of the quantum behavior. (b) The stimulated Raman kick and RF pulse used to implement the quantum walk may also contribute to the internal heating and decoherence of the atoms after a few iterations of the walk.

The measurement scheme proposed by Chandrashekar (2006) involves collapsing the final BEC superposition in the position space by applying multiple microtraps with spacing $l$ between the potential wells, while removing the dipole trap potential. Microtraps are created by illuminating a set of microlenses with a red-detuned laser beam, such that a single dipole trap is formed in each of the foci of the individual microlenses (Dumke et al. 2002). A fluorescence measurement on the condensate in the microtrap can identify the final position of the BEC.

The use of optical microtraps in quantum walks can in fact be traced back to an earlier work by Eckert et al. (2005) in which they proposed implementing the quantum walk itself (not just the measurement) using the controlled motion of a single neutral atom in arrays of microtraps. The concept, which can be readily applied to neutral atoms trapped in optical lattices or even magnetic potentials, involves the use of spatially delocalized qubits implemented by the presence of the atom in the ground state of one of two adjacent trapping potentials (Mompart et al. 2003). Hence not only the walking (conditional translation) can take place in the position space, the quantum coin can also be represented by a spatial degree of freedom. In contrast to proposals based on optical lattices (Sect. 3.6.2), no state-dependent potentials are necessary.

Eckert et al. (2005) employed two sets of microtraps generated by illuminating a set of microlenses with two independent laser beams. By changing the angle between the two lasers, it is possible to spatially move the two sets of traps relative to one another, allowing the atom to propagate between different microtraps. The optical potential at each trapping site has a Gaussian shape given by

$$V(x) = -\mathcal{V}\exp\left(-\frac{1}{2\mathcal{V}}m\omega_x^2 x^2\right) = -\mathcal{V}\exp\left(-\frac{\hbar\omega_x}{2\mathcal{V}}(\alpha x)^2\right), \tag{3.130}$$

where $\alpha^{-1} = \sqrt{\hbar/m\omega_x}$ denotes the spread of the ground state in position space, with $m$ being the atomic mass. Using $\mathcal{V} = 200\,\hbar\omega_x$ ensures that the traps are deep enough to be described by harmonic potentials of frequency $\omega_x$ in the limit of large separation. The sites along an array of microtraps are labeled with $k = 0, \pm1, \pm2\cdots$ and pairs of microtraps are used to implement the coin basis states $|+\rangle$ and $|-\rangle$ which are defined by a single atom occupying the ground state of one of the two adjacent traps. Unitary operations are performed by approaching the two traps forming the coin, allowing the atom to tunnel between them.

The authors presented two configurations for implementing such a quantum walk. In the first configuration the two independent sets of microtraps are arranged to form two adjacent rows. Figure 3.58 shows the first step in the temporal evolution of this configuration with the corresponding atomic probability distributions overlayed as a density plot. Each coin is defined by a pair of microtraps, one trap from each row. By moving both rows in opposite directions at an optimized distance and velocity, the coin operations $\hat{C}$ are performed when the traps pass each other at close distance. The trap sites $k$ (when the two rows are aligned) represent the alternate position states of the walk: even position states $|i\rangle \equiv |2k\rangle$ for even steps and odd position states $|i\rangle \equiv |2k+1\rangle$ for odd steps of the walk. This is convenient since the quantum walk procedure guarantees that every other position state would have a zero probability amplitude. The conditional translation $\hat{T}$ is implicit through a redefinition of the qubits after each move, as depicted in Fig. 3.58(IV). In this way

$$\hat{T}|i\rangle|-\rangle = |i-1\rangle|-\rangle, \quad\text{and} \tag{3.131}$$

$$\hat{T}|i\rangle|+\rangle = |i+1\rangle|+\rangle, \tag{3.132}$$

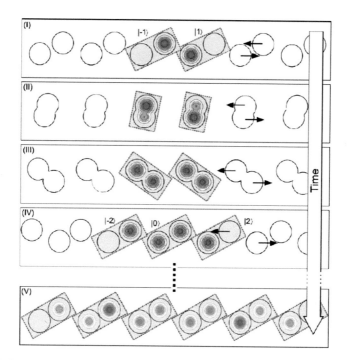

**Fig. 3.58** Configuration with two rows of traps; the qubit is implemented "perpendicular" to the rows (*dashed rectangles* show which two traps form each qubit). The upper (lower) row moves to the left (right) with constant velocity (see *arrows*). (*I*) After the first step $|\psi\rangle = 1/\sqrt{2}(|-1,-\rangle + |+1,+\rangle)$. (*II*) and (*III*) The coin operation, in this case a Hadamard gate, is performed when the traps pass each other. (*IV*) The translation operator is implicit through a redefinition of the qubits. After an even (odd) number of shift operations only the even (odd) qubits are defined (compared to the standard quantum walk definition). (*V*) The probability distribution after the sixth displacement operation. For the numerical simulation the authors used a potential that, along the line connecting the centers of two traps, reads $V(x) = \hbar\omega_x \min\{\alpha^2(x-d)^2, \alpha^2(x+d)^2\}$. The velocity is chosen such that during the passing of two traps a Hadamard operation is performed. The initial state is $|\psi_{\text{init}}\rangle = 1/\sqrt{2}(|0,-\rangle + |0,+\rangle)$ (From Eckert et al. (2005))

as required. This implementation can become quite challenging however if continuous displacement of the microtrap rows requires mechanical movement of an array of lenses. To overcome this problem, Eckert et al. (2005) conjectured using holographic techniques for generating the arrays of microtraps (Bergamini et al. 2004).

In the second configuration the coin basis states are given by the adjacent sites along a single row of microtraps and the position states $|i\rangle$ are defined by the $i$th block representing their respective pair of coin states $|+\rangle \equiv |2k\rangle$ and $|-\rangle \equiv |2k+1\rangle$. Figure 3.59 illustrates this configuration. For coin operations, the traps $2k$ and $2k+1$ are brought within interaction proximity, while for the conditional translation, a swap operation between traps $2k+1$ and $2(k+1)$ moves

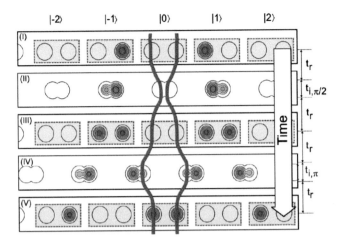

**Fig. 3.59** Configuration with a single row of traps; the qubit is implemented parallel to the rows (*gray boxes*). (*I*) After the first step $|\psi\rangle = 1/\sqrt{2}(|-1,+\rangle + |+1,-\rangle)$; (*II*) and (*III*) traps inside each qubit are approached to produce the coin operation; (*IV*) and (*V*) the translation operator is realized through approaching traps of adjacent qubits (From Eckert et al. (2005))

the quantum walker one step in the appropriate direction, but has the side effect of flipping the coin states upon each move, i.e.

$$\hat{T}|i\rangle|-\rangle = |i-1\rangle|+\rangle, \quad \text{and} \tag{3.133}$$

$$\hat{T}|i\rangle|+\rangle = |i+1\rangle|-\rangle. \tag{3.134}$$

This flip-flop walk (Ambainis et al. 2005) can be corrected by applying a second swap operation between traps $2k$ and $2k+1$ which can be built into the coin operator.

The main experimental requirement of this proposal is to be able to move all odd or all even traps as a whole to both directions, thus approaching each second trap to its left or right neighbor. Initially the atom is prepared in the supper position state $|\psi_{\text{init}}\rangle = 1/\sqrt{2}(|+\rangle + |-\rangle)$. The traps are then adiabatically moved between the maximal and minimal distances $d_{\max}$ and $d_{\min}$, where $\alpha d_{\max} = 60$ and $\alpha d_{\min} = 28.8$. The movement of the traps are carefully optimized while suppressing transitions between motional states. In this way, for a typical trapping frequency of about $\omega_x = 10^5 \, \text{s}^{-1}$, the time $t_r$ necessary to approach or separate the traps can be reduced to $\omega_x t_r = 100$ or below while maintaining a fidelity larger than 0.999. The time $t_i$ for which the distance between the traps is kept constant at $d_{\min}$ is chosen such that alternately a swap or coin operation is applied. Consequently each step of the walk is estimated to take around 5 ms. Figure 3.60(I) shows the probability distribution across the traps after $n = 10, 20,$ and 25 steps obtained from the integration of the 1D Schrödinger equation inside each microtrap.

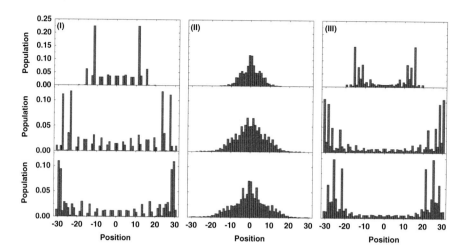

**Fig. 3.60** The probability distribution to find the atom at a specific trap site, with (*I*) the ground state and (*II*) the first excited state as the initial vibrational state, for a 1D quantum walk on a finite line of 62 traps; from *top* to *bottom* distributions after $t = 10, 20$, and 30 steps are shown. Parameters $V = 200\,\hbar\omega_x$, $\alpha d_{\max} = 60$, $\alpha d_{\min} = 28.8$, and $\omega_x t_r = 100$; $\omega_x t_{i,\pi} = 20.25$ for the $\pi$ pulse and $\omega_x t_{i,\pi/2} = 112$ for the $\pi/2$ pulse. (*III*) Like (*II*), but with $\omega_x t_r = 200$, such that non-adiabatic excitations are suppressed (From Eckert et al. (2005))

Eckert et al. (2005) pointed to two particular sources of experimental imperfection. One is the impact of the internal state of the atom (ground or excited) on the precise timing of the changes in the trap separation, which the tunneling process as well as the adiabaticity crucially depend on. For simulations presented in Fig. 3.60(I) the durations $t_r$ and $t_i$ are chosen to apply the correct operations only for the vibrational ground state. However, a simulation for an atom initially in the first excited vibrational state (Fig. 3.60(II)), shows a markedly different distribution. A more realistic assumption than starting from a pure state with the atom being in a specific vibrational level is to consider a thermal Boltzmann distribution of the vibrational modes given by the density operator

$$\rho = \frac{1}{z} \sum_{j=0}^{\infty} e^{-\beta E_j} |j\rangle\langle j|, \qquad (3.135)$$

where $z = \sum_{j}^{\infty} e^{-\beta E_j}$, $\beta = 1/k_B T$ and $E_j$ is the energy of the $j$th vibrational mode. In this case the experimentally accessible probability distributions are the classically averaged probability distributions, weighted with factors $\exp(-\beta E_j)/z$. The respective probability distributions after $n = 20$ steps are shown in Fig. 3.61 for initial ground-state populations of 50 and 25 %, corresponding to a temperature of $T = 1.1$ and $2.7\,\mu K$ (for Rb atoms) respectively.

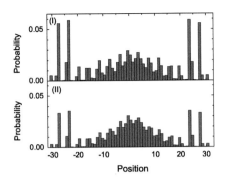

**Fig. 3.61** Probability distributions after n = 20 steps for thermal Boltzmann distributions of vibrational modes; initial ground-state population (*I*) 50 % and (*II*) 25 %. All other parameters as in Fig. 3.3 (From Eckert et al. (2005))

The second experimental imperfection stems from the external noise, causing the vibration of microtraps. This is modeled by introducing harmonic oscillations (at a fixed frequency far above the trapping frequency), with the assumption that the movement of even and odd traps is correlated. The consequence of this random variation is similar to the effects observed in the presence of decoherence and the final probability distribution approaches a Gaussian.

A naturally dominant decoherence mechanism in this system is expected to be the scattering of photons from the trapping laser, with scattering rates in the order of $0.1–1 \, \text{s}^{-1}$. Hence the probability for a decoherence event to occur within a single step (around 5 ms) is in the range $p = 0.0005$ to $p = 0.005$. Eckert et al. (2005) estimated the crossover from the quantum to the classical distribution to take place at $np \approx 2.6$, after $n$ steps of the walk. For this reason it should be possible to observe the quantum walk in such systems and to analyze changes caused by temperature and external vibration without being limited by decoherence from photon scattering.

Eckert et al. (2005) also proposed a scheme for realizing a 2D quantum walk on a regular square optical lattice using a 4D coin with basis states $|+, 0\rangle \equiv |\nearrow\rangle$, $|-, 0\rangle \equiv |\nwarrow\rangle$, $|-, 1\rangle \equiv |\swarrow\rangle$ and $|+, 1\rangle \equiv |\searrow\rangle$, to control the displacement of the particle into four possible directions. The authors implemented such a coin by combining the spatial qubit states ($|+\rangle$ and $|-\rangle$) with the internal hyperfine states of the atom ($|0\rangle$ and $|1\rangle$) defined in Eq. 3.125). Unitary coin operations involve approaching the traps as before, as well as putting the atom in a superposition of the two hyperfine levels using a Raman transition or a microwave pulse. A combination of the population swap by approaching the traps as well as spin-dependent transport (similar to the proposal of Dur et al. (2002)) is used to implement the conditional translation operator.

In an entirely different approach Li et al. (2008) proposed a scheme involving the passage of light through a series of cells each trapping a coherent gas of cold alkali atoms such as $^{85}$Rb. The coin basis states of the walk are spanned by the circular polarization states $|\sigma^-\rangle$ and $|\sigma^+\rangle$ of the incident light while the position states are encoded by the incremental phase shifts $\pm\delta\phi$ picked up by each polarization state as light passes through the atomic cells.

Figure 3.62 presents a schematic diagram of the proposed setup where at each step $N$ of the walk the probe field $\mathbf{E}_p$ passes through a black box optical

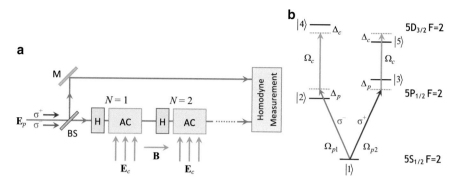

**Fig. 3.62** (**a**) Proposed experimental apparatus for performing a one-dimensional quantum walk. BS is a beam splitter, M is a high reflective mirror and each step $N$ of the walk is performed via the passage of light through the Hadamard gate H and an atomic cell AC (Adapted from Li et al. (2008)). (**b**) Energy level diagram and excitation scheme of a lifetime-broadened five-level atomic system that interacts with a weak pulsed probe field of Rabi frequency $\Omega_p$ and a strong continuous-wave coupling field of Rabi frequency $\Omega_c$; $\Delta_p$ and $\Delta_c$ are relevant one- and two-photon detunings. The linearly polarized probe field can be considered as a superposition of the *right* ($\sigma^+$) and *left* ($\sigma^-$) circularly polarized components with the Rabi frequencies $\Omega_{p1}$ and $\Omega_{p2}$, respectively

unit H performing the Hadamard coin operation followed by the atomic cell AC responsible for the conditional phase shift. The probe field is linearly polarized and tuned to the $5S_{1/2} \longleftrightarrow 5P_{1/2}$ transition, propagating along the direction of an external magnetic field **B** responsible for the Zeeman splitting of the atoms trapped by each cell (see Appendix E.2.1).

A key feature of this proposal is a process known as electromagnetically induced transparency (EIT) involving the application of a strong secondary laser beam known as the coupling field $\mathbf{E}_c$ which is tuned to the $5P_{1/2} \longleftrightarrow 5D_{3/2}$ transition and its propagation direction is perpendicular to the external magnetic field. The essential feature of an EIT-based system is that the absorbtion of the probe field can be largely suppressed and hence an initially opaque optical medium becomes highly transparent, reaching transmission ratios as high as 99.5 %. In comparison, for a typical quartz wave plate with refractive index $n = 1.54$, the transmission ratio is lower at 95.5 %. This, the authors note, is a significant advantage of this proposal over those involving standard optical elements such as wave plates, particularly for a quantum walk with a sizable number of steps.

The probe field's linear polarized can be expressed as a superposition of the circular polarization states $|\sigma^-\rangle$ and $|\sigma^+\rangle$ on which the Hadamard coin operator H acts. Li et al. (2008) presented only a brief sketch for the implementation of such an operator, referring to a configuration proposed by Cerf et al. (1998) involving a lossless symmetric beam splitter placed between two $-\pi/2$ phase shifters. A closer examination of the work of Cerf et al. (1998) however reveals that their proposed Hadamard gate operates on spatial modes of light (horizontally and vertically propagating beams), and not the polarization states. An adaptation this optical gate to accept polarization input states would necessitate the use of additional polarizing beam splitters. Alternatively a single wave plate can implement the Hadamard and other coin operations, a technique employed by numerous optical implementations we have already examined. It should be noted however that in both cases the introduction of various optical elements would negatively impact the transmission efficiency of the overall system touted as a key advantage of this scheme.

The implementation of the conditional translation takes advantage of the Zeeman splitting due to the external magnetic field which gives rise to two possible transitions $|1\rangle \longleftrightarrow |2\rangle$ and $|1\rangle \longleftrightarrow |3\rangle$ involving the absorption of $-\hbar$ and $+\hbar$ units of angular momentum respectively. These transitions can therefore be conveniently driven by the $\sigma^-$ and $\sigma^+$ polarization components, as depicted in Fig. 3.62b, while the coupling field drives the transitions $|2\rangle \longleftrightarrow |4\rangle$ and $|3\rangle \longleftrightarrow |5\rangle$, together forming a ladder configuration. Now considering the complex valued electric susceptibilities $\chi_\pm$ for the $\sigma^\pm$ components of the probe field we find

$$\chi_- = -\frac{N_a \mathbf{d}_{SP}}{\hbar \epsilon_0} \frac{\rho_{12}}{\Omega_{p1}}, \quad \text{and}$$

$$\chi_+ = -\frac{N_a \mathbf{d}_{SP}}{\hbar \epsilon_0} \frac{\rho_{13}}{\Omega_{p2}}, \tag{3.136}$$

where $N_a$ is the density of the atomic gas, $\mathbf{d}_{SP}$ is the electric dipole matrix element associated with the 5S $\longleftrightarrow$ 5P transition, $\epsilon_0$ is the vacuum dielectric constant, $\Omega_{p1}$ and $\Omega_{p2}$ are the probe field Rabi frequencies associated with transitions $|1\rangle \longleftrightarrow |2\rangle$ and $|1\rangle \longleftrightarrow |3\rangle$ respectively and $\rho_{ij}(t) \equiv \langle i|\psi(t)\rangle \langle \psi(t)|j\rangle$ are components of the density matrix for the atom with wave-function $\psi(t)$. Expressions for $\rho_{12}$ and $\rho_{13}$ in the linear regime can be readily obtained by solving the density matrix equations for the complete system, assuming probe field intensities much weaker than that of the coupling field, and using the Fourier transform technique used by Wu and Deng (2004). Presenting the solution, Li et al. (2008) showed that the two circular components $\sigma^\pm$ of the probe field have different dispersion properties determined by Re $[\chi_\pm]$ and hence acquire different phase shift modifications when passing through the atomic cells. More specifically, probe photons with frequency $\omega_p$ propagating through the atomic medium of length $L$ undergo a transformation

$$|\sigma^\pm\rangle \longrightarrow \exp(-i\phi_\pm)|\sigma^\pm\rangle, \tag{3.137}$$

where $\phi_\pm = \omega_p \bar{n}_\pm L/c$ and $\bar{n}_\pm$ is the real part of the refractive index $n_\pm = \sqrt{1 + \chi_\pm} \simeq 1 + \chi_\pm/2$. Hence $\phi_\pm \simeq \phi_0 + \delta\phi_\pm$ where $\phi_0 = \omega_p L/c$ and the phase shift modification $\delta\phi_\pm = \omega_p \text{Re}[\chi_\pm] L/2c$. The authors noted that under a realistic set of parameters it is possible to satisfy the condition $\delta\phi_+ = -\delta\phi_- = \delta\phi$, although it is not strictly necessary (see the implementation of Schreiber et al. (2010) for non-symmetric steps). The quantum walk position states at the $N$th step of the walk can now be formally encoded as $|\phi_k\rangle_N \equiv |N\phi_0 + k\delta\phi\rangle$ for $k = \pm 0, 1, 2 \ldots N$, where $\delta\phi$ is the quantum walk step length and we may drop the global phase shift $N\phi_0$ as it only represents a constant shift in the entire position space of the walk without altering in any way the probability distribution outcome. Using this formalism, the quantum walker's passage through the atomic cell can be described by the transformation

$$|\sigma^\pm, \phi_k\rangle \xrightarrow{\text{AC}} |\sigma^\pm, \phi_{k\pm1}\rangle, \qquad (3.138)$$

which is readily recognizable as the required conditional translation, and the entire position space has undergone an additional but unimportant global phase shift $\phi_0$. A complete $N$ step quantum walk can now be realised by allowing the probe filed to pass through $N$ identical atomic cells before constructing the final probability distribution for the relative phase shifts. To this end, the authors proposed a homodyne detection technique where modulations in the output phase are detected via interference with an unaltered reference field.

### 3.6.2 Optical Lattices

Dur et al. (2002) proposed implementing a quantum walk on a line or on a circle using a single neutral atom confined to a periodic optical potential known as an optical lattice (see Appendix E). This can be accomplished by first loading a Bose-Einstein Condensate (BEC) of the atoms into an optical lattice, then making a transition to the Mott insulator state with a filling factor of 1 (see Appendix E.1) and finally, selectively addressing and depleting the unneeded lattice sites (Würtz et al. 2009) (although at the time, this final step would have posed a formidable experimental challenge).

The position states $|x\rangle$ of the walk are represented by positions $x$ of the individual lattice sites in a 1D optical lattice and the coin basis states of the walk are taken to be the hyperfine states of the atom (Eq. 3.125). The production of the optical lattice naturally leads to a pair of superimposed lattices, labeled 0 and 1, which respectively trap the atomic wave-function in the internal (or spin) state $|0\rangle$ and $|1\rangle$ (see Appendix E.3). Dur et al. (2002) exploited the spin dependant lattice movements (see Fig. E.7) to implement the conditional translation operator, while laser pulses are used to implement the coin operator by manipulating the internal state of the atom via Raman transitions (see Appendix B.3.1). The authors first considered the case where lattice 0 (trapping the internal state $|0\rangle$) moves with constant velocity $v_0 = -v$ to the left, while lattice 1 (trapping the internal state $|1\rangle$) moves with a constant velocity $v_1 = v$ to the right. This trivially performs the translation operation given by

$$\hat{T}|x\rangle|0\rangle = |x - d\rangle|0\rangle, \quad \text{and} \qquad (3.139)$$

$$\hat{T}|x\rangle|1\rangle = |x + d\rangle|1\rangle, \qquad (3.140)$$

where $d = \lambda/2 \approx 425\,\text{nm}$ is the lattice period. Dur et al. (2002) proposed using standard Raman pulse or microwave techniques to realize the coin operator $\hat{C}$ (e.g. Hadamard rotation) by using fast laser pulses to achieve a coherent superposition of two internal states (see Appendix C). The coin operations are applied to all lattices simultaneously using a non-focused laser beam. This is convenient since performing individual operations on neighbouring lattice sites can be challenging, since the lattice period is limited by the optical wavelength which is smaller than the best achievable focusing width of the laser beams $\approx 1\,\mu\text{m}$ (for ways to achieve single site addressability, see the work of Manouchehri and Wang (2009) and references therein). The continuous motion of the lattices is broken into discrete time steps, with the $n$th step of the quantum walk completed at time $t_n = nd/v$ when the two moving lattices are again on top of each other. The coin operation is performed at this critical moment assuming that the operation time is a fraction of the lattice motional period. The relative motion of the lattices must also be sufficiently slow such that the internal state of the atom remain stable and the particle remains trapped in the potential throughout the procedure. This corresponds to a maximum number of about $n = 10^4$ time steps, assuming $100\,\mu\text{s} \lesssim t_1 \lesssim 1\,\text{ms}$, and given the spontaneous emission lifetime of the atom in the lattice, which is of the order of several seconds. The final atomic distribution within the lattice can be readily obtained via a fluorescence measurement, together with several repetitions of the experiment (for a review of various measurement techniques see Gericke et al. (2008) and references therein). In practice, the measurement efficiency can be improved by conducting the experiment in a 2D or 3D optical lattice, whereby numerous quantum walks can take place simultaneously in parallel rows.

There are a number of challenges however arising from the above implementation. As already mentioned, the laser pulses implementing the coin operation have to be fast compared to the time scale of the lattice movement. In addition, changing the internal state of the atom implies a sudden momentum change. In going from state $|0\rangle$ to $|1\rangle$ for example, the atom is no longer trapped in the left-moving lattice but is suddenly in the right-moving one. This momentum change may lead to heating of the atom and it may eventually even escape from the trap. To overcome these issues, Dur et al. (2002) introduced an improved mechanism whereby, instead of moving the lattices with constant velocity, harmonic oscillations around the central position $x_0 = 0$ facilitate the spin dependant transport. Choosing an oscillation period in the range $100\,\mu\text{s}$–$1\,\text{ms}$ ensures that the adiabaticity requirement for the lattice movement is well fulfilled (i.e. the atom remains in the motional ground state throughout the procedure). The internal state of the atom is manipulated at the moment when the two counter-oscillating lattices are exactly on top of one another and the atom velocity is zero. One could in principle also stop the lattice movement at these times until the internal state transformation has been completed. This would effectively remove the requirement that the coin operation should be fast compared to the time scale of the lattice movement. Since in this scheme the direction of lattice motion changes after every step, for the walk to precede forward a swap operation $\hat{\sigma}_x$ is performed at every time step $t_n$. This introduces an additional subtlety involving the order in which the operators $\hat{\sigma}_x$ and $\hat{C}$ act on the system. The coin operator is

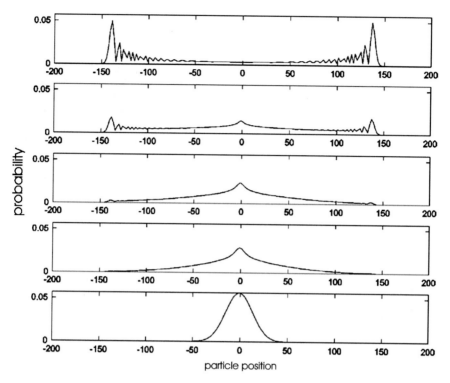

**Fig. 3.63** Probability distribution after $n = 200$ steps of quantum walk on a line with imperfect operations using error model 1 for $p = 1, 0.99, 0.97, 0.95, 0$ from *top* to *bottom*. Only even positions are plotted, since odd positions are not occupied. The lowest curve corresponds to the probability distribution of the classical random walk on the line (From Dur et al. (2002))

expected to act on the atomic wave-function when state $|0\rangle$ is trapped by lattice 0, and $|1\rangle$ by lattice 1. Hence the ordering of the two operators is switched at every step. In other words at times $t_n$ when $n$ is even, the effective coin operator is $\hat{\sigma}_x \hat{C}$ and when $n$ is odd, the effective coin operator is $\hat{C}\hat{\sigma}_x$.

Irrespective of the mechanism adopted for implementing the walk, there remain different kinds of errors which may influence the ideal evolution. These include errors in lattice movements which may lead to motional excitations of the atom. The internal state of the atom is also influenced by the decoherence resulting from uncontrollable phase shifts and imperfections in the manipulation by means of laser pulses, as well as fluctuations in the trapping potential during lattice shifts. Dur et al. (2002) attempted to model this (error model 1) by assuming that the desired coin operation $\hat{C}$ is performed with probability $p$ at each time step, while with probability $1 - p$, a completely depolarized, random state is produced. The resulting probability distribution is plotted in Fig. 3.63 where, the collapse of the ideal quantum walk distribution to a classical Gaussian distribution is clearly visible. Errors may also accumulate in the position space of the atom, introduced by tunneling of atoms between neighboring lattice sites. In this error model (error

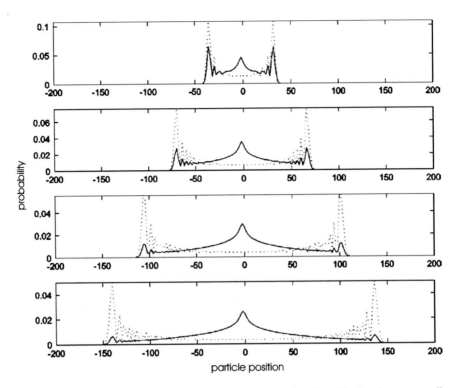

**Fig. 3.64** Probability distribution for ideal quantum walk on a line (*dotted line*) and quantum walk with imperfect operations using error model 2 for p = 0.98 (*solid line*) after $n = 50, 100, 150, 200$ steps from *top* to *bottom*. Only even positions are plotted, since odd positions are not occupied (From Dur et al. (2002))

model 2) the ideal evolution occurs with probability $q$, while with probability $(1-q)$, the atom tunnels to one of its neighboring lattice sites. The probability distribution resulting from this evolution is depicted in Fig. 3.64. Figure 3.65 shows how the probability of detecting the walker (neutral atom) ten lattice sites away from the origin changes between an ideal quantum walk, an imperfect quantum walk (using error model 1) and a classical random walk.

Karski et al. (2009) later demonstrated a physical realization of Dur et al.'s proposal (2002). They utilized a single Cs atom, trapped in the potential wells of a 1D optical lattice with site spacing of 433 nm. Coin basis states of the walk are represented by the hyperfine states $|0\rangle \equiv |F = 4, m_F = 4\rangle$ and $|1\rangle \equiv |F = 3, m_F = 3\rangle$ of the atom, coherently coupled via a resonant microwave radiation around 9.2 GHz which enables the implementation of Hadamard-type coin operations. The conditional translation operator is performed using the spin(state)-dependant lattice transport (see Appendix E.3), moving the spin state $|0\rangle$ adiabatically to the right and $|1\rangle$ to the left along the lattice axis within 19 $\mu$s. To increase the fidelity of the walk spin-echo operations are combined with each coin operation, leading to a coherence time of 0.8 ms.

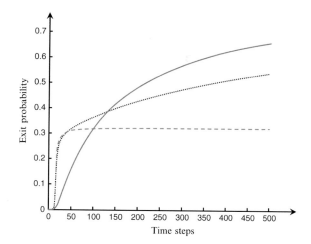

**Fig. 3.65** Bounded quantum walk with barrier at position $x = -10$. Probability that the atom was observed at the barrier plotted as a function of the number of steps for classical random walk (*solid line*), ideal quantum walk (*dashed line*), and quantum walk with imperfections in the manipulation of the internal state of the atom (*dotted line*), using error model 1 and $p = 0.99$ (From Dur et al. (2002))

After $N$ steps of the quantum walk, the final atom distribution is probed by fluorescence imaging. From these images, the exact lattice site of the atom after the walk is extracted. The final probability distribution to find an atom at lattice site $x$ after $N$ steps is obtained from the distance each atom has walked by taking the ensemble average over several hundreds of identical realizations of the sequence.

Karski et al. (2009) demonstrated a quantum walk with $N = 6$ steps with the resulting probability distribution shown in Fig. 3.66. This is contrasted with the binomial distribution of a (classical) random walk, recovered by omitting the spin-echo from the coin operation and additionally waiting 400 ms between each coin and its subsequent shift operation, thus destroying the phase relation between subsequent steps of the walk. Karski et al. (2009) then investigated the scaling of the quantum walk width with the number of steps and reported that it follows closely the expected linear behavior for up to ten steps, as compared with the typical square-root scaling of a purely classical random walk (Fig. 3.67).

To demonstrate the spatial coherence of the state over all populated lattice sites following a six-step quantum walk sequence, Karski et al. (2009) inverted the coin and shift operations, and continued the walk for six additional steps, depicted in Fig. 3.68. Ideally, the inversion acts as an effective time-reversal and refocuses the multipath interference pattern of the wave-function back to the initial lattice site. The authors found partial refocusing of 30 % of the atomic population to the expected lattice site reflecting the fraction of atoms which have maintained coherence throughout the sequence.

An alternative quantum walk scheme proposed by Joo et al. (2007) was based on a single atom trapped inside a time-modulated state-dependant *super*-lattice (Peil et al. 2003; Kay and Pachos 2004; Calarco et al. 2004) formed by interfering two optical lattices, each of which is produced by a pair of laser beams intersecting at an angle $\Theta_i \leqslant \pi$ $(i = 1, 2)$ depicted in Fig. 3.69. The resulting optical lattices are aligned with the $z$-axis and have an effective optical period $d_i = \lambda/[2 \sin(\Theta_i/2)]$ which reduces to the expected $\lambda/2$ for exactly counter propagating lasers. Therefore

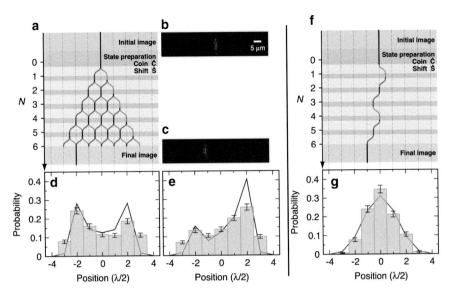

**Fig. 3.66** (**a**) Schematic experimental sequence for the quantum walk showing the paths for the internal states $|0\rangle$ (*green*) and $|1\rangle$ (*red*). The walking distance is extracted from the initial (**b**) and final (**c**) fluorescence image. The results of several hundreds of identical realizations form the probability distribution, which is symmetric for the initial state $(|0\rangle + |1\rangle)/\sqrt{2}$ (**d**) and antisymmetric for the initial state $|1\rangle$ (**e**). The analogous classical random walk sequence (**f**) yields a binomial probability distribution (**g**). The displayed path is one of many random paths that the atom can take. Measured data are shown as a histogram, and the theoretical expectation for the ideal case is denoted with a *solid line*. Error bars indicate the statistical $\pm 1\sigma$ uncertainty (From Karski et al. (2009))

changing the angle $\Theta_i$ makes it possible to create two parallel optical lattices with differing periods. Consequently the superposition of the two lattices (see Eq. E.15) produces a super-lattice with a versatile potential profile, given by

$$\mathcal{V}_{\pm 1/2}(z) = \sum_{i=1}^{2} \frac{1}{2} V_{\mathrm{max},i} \cos(2k_i z + \phi_i \pm \theta_i) + \frac{1}{6} V_{\mathrm{max},i} \cos(2k_i z + \phi_i \mp \theta_i) + \frac{2}{3} V_{\mathrm{max},i}, \tag{3.141}$$

where $k_i = 2\pi/d_i$ is the $i$th lattice wave number, $\phi_i$ and $\theta_i$ represent the polarization and phase angles between the two beams forming lattice $i$ and $V_{\mathrm{max},i}$ is the unit light shift which depends on the laser intensity and detuning.

In their proposal Joo et al. (2007) employed a pair of lattice potentials with parameters $d_1 = 2\lambda$, $d_2 = 4\lambda$, $\theta_1 = 0$, $\theta_2 = \pi/2$ and $\phi_1 = 0$, resulting in the super-lattice potential

$$\mathcal{V}_{\pm 1/2}(z) = \mathcal{V}_1 \cos(\frac{2\pi}{\lambda} z) + \mathcal{V}_2 \cos(\frac{\pi}{\lambda} z + \phi_2 \pm \frac{\pi}{2}), \tag{3.142}$$

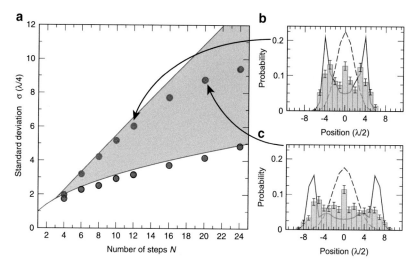

**Fig. 3.67** (**a**) Scaling of the SD of the measured spatial probability distributions for quantum walk (*red*) and random walk (*green*). The *solid lines* indicate the expectations for the ideal cases. Error bars are smaller than the size of symbols. The measured quantum walks follow the ideal linear behavior until, because of decoherence, they gradually turn into a random walk. The probability distributions for $N = 12$ (**b**) and $N = 20$ (**c**) show a gradual change from the quantum to a classical shape. The theoretical prediction is shown as a *solid line* for the pure quantum walk and as a *dashed line* for the random walk (From Karski et al. (2009))

where $\mathcal{V}_1 = 4V_{\mathrm{max},1}/3$ and $\mathcal{V}_2 = V_{\mathrm{max},2}/3$ are the potential amplitudes and the unimportant global potential shift is neglected. Initially the second optical lattice is switched off, leaving a regular lattice given by

$$\mathcal{V}_{\pm 1/2}(z) = \mathcal{V}_1 \cos(\frac{2\pi}{\lambda}z). \tag{3.143}$$

A single neutral $^{87}$Rb atom is then trapped inside one of the potential minima and the position and coin states of the walk are defined similar to the scheme of Dur et al. (2002); the position states $|x\rangle$ are the minima of the optical potential and the coin basis states $|0\rangle$ and $|1\rangle$ are the hyperfine states of the atom (Eq. 3.125). The innovation involved here is in the walking procedure itself or more precisely the implementation of the state dependent translation operator $\hat{T}$. It involves switching on the second lattice for a duration $t$ and alternating $\phi_2$ between $\pi/2$ and $-\pi/2$ every time the lattice is switched on. The resulting supper-lattice optical potential is given by

$$\mathcal{V}^+_{\pm 1/2}(z) = \mathcal{V}_1 \cos(\frac{2\pi}{\lambda}z) \pm \mathcal{V}_2 \cos(\frac{\pi}{\lambda}z) \tag{3.144}$$

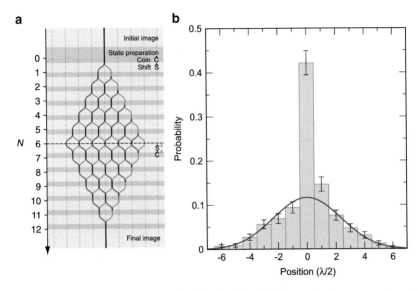

**Fig. 3.68** Time-reversal sequence for refocusing the delocalized state of a six-step quantum walk. (**a**) After six steps, the total application of the coin $\hat{C}$ and shift $\hat{S}$ operator is reversed, where $(\hat{S}\hat{C})^{-1} = \hat{C}^{-1}\hat{S}^{-1}$. (**b**) The resulting probability distribution shows a pronounced peak at the center, to where, ideally, the amplitude should be fully refocused. A refocused amplitude of 30 % is observed, surrounded by a Gaussian background (fitted curve) (From Karski et al. (2009))

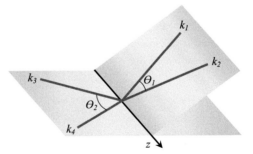

**Fig. 3.69** Laser beam configuration for producing an optical super-lattice. The plane containing the wave vectors $\vec{k}_{1,2}$ of the first laser pair intersects the plane containing the wave vectors $\vec{k}_{3,4}$ of the second laser pair at the $z$-axis. The planes can have a finite angle between them. The angle between the lasers in each of the pairs are $\Theta_1$ and $\Theta_2$, respectively

for $\phi_2 = -\pi/2$ and

$$V_{\pm 1/2}^{-}(z) = V_1 \cos(\frac{2\pi}{\lambda}z) \mp V_2 \cos(\frac{\pi}{\lambda}z) \tag{3.145}$$

for $\phi_2 = \pi/2$. This has the effect of selectively lowering and raising the potential barrier between adjacent lattice sites, depending on the state of the atom. In other

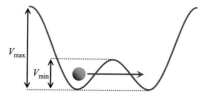

**Fig. 3.70** Tunneling of an atom in the internal state $|0\rangle$ from the *left site* to the *right site* in a double well with $V_{max} = V_1 + V_2$ and $V_{min} = V_1 - V_2$ (Adapted from Joo et al. (2007))

words, when the supper-lattice potential is $V^+$, atoms in the $m_s = 1/2$ or $|0\rangle$ state perceive that the odd barriers are raised to $V_{max} = V_1 + V_2$ and even barriers are lowered to $V_{min} = V_1 - V_2$ as illustrated in Fig. 3.70. Likewise atoms in the $m_s = -1/2$ or $|1\rangle$ state perceive the opposite potential shifts. When the supper-lattice potential is switched to $V^-$, the ordering is reversed.

This mechanism can be used to achieve a state dependant control of the tunneling rates between the adjacent lattice sites, which forms the basis of Joo et al.'s proposed scheme (2007). Assuming an ideal case where $V_{max}$ produces no tunneling and $V_{min}$ enables a tunneling between two positions in the double well, one can describe the tunneling process between the adjacent wells $|x_1\rangle$ and $|x_2\rangle$ using the Hamiltonian

$$\hat{H} = -J(a_1^\dagger a_2 + a_2^\dagger a_1), \tag{3.146}$$

where $J$ is the tunneling interaction. The resulting time evolution operator is described by

$$U(t) = \cos(\frac{1}{2}Jt)\mathbb{1} + i\,\sin(\frac{1}{2}Jt)\sigma_x, \tag{3.147}$$

where $t$ is the evolution time and $\sigma_x$ is one of the Pauli operators. Hence in order to obtain perfect tunneling between the two sites, a time $t = \pi/J$ is required after which the initial amplitude in position state $|x_1\rangle$ is entirely transferred to state $|x_2\rangle$. The tunneling parameter $J$ is dependant on the barrier height $V$ according to

$$J(V) = \frac{E_R}{2} \exp\left(-\frac{\pi^2}{4}\sqrt{\frac{V}{E_R}}\right)\left[\sqrt{\frac{V}{E_R}} + \left(\frac{V}{E_R}\right)^{3/2}\right], \tag{3.148}$$

where $E_R = \hbar^2 k^2/(2m)$ is the recoil energy and $m$ is the atomic mass. Hence the requirement for performing the quantum walk is to make the ratio $J(V_{max})/J(V_{min}) \equiv J_{max}/J_{min}$ sufficiently small in order to suppress undesirable tunneling.

With the tunneling parameters optimized, the quantum walk can proceed as depicted in Fig. 3.71. Starting with the atom in a mixed internal state ($\psi_0 = 1/\sqrt{2}(|0\rangle + i|1\rangle)$), loaded inside the lattice minimum at $x = 0$, the second optical

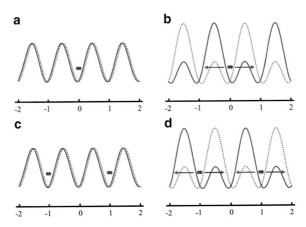

**Fig. 3.71** (a) The atom with internal states $|0\rangle$ and $|1\rangle$, represented by colors *blue* and *red*, is initially located at site $x = 0$ of a single optical lattice. (b) By turning on the second optical lattice, state dependant tunneling is activated from even sites in the super-lattice to their neighboring sites. (c) A Hadamard operation is performed in the optical lattices. (d) By altering the super-lattice potential, state dependant tunneling is activated from odd sites in the super-lattice to their neighboring sites (Adapted from Joo et al. (2007))

lattice is switched on to produce the optical potential $\mathcal{V}^+$. After a time $t = \pi/J_{\min}$ both internal states of the atom have been translated to neighboring sites and the second lattice is switched off. The coin operator $\hat{C}$ is then applied by rotating the internal states of the atom using standard Raman pulse or microwave techniques (see Appendix C). To perform the next translation the second lattice is again switched on, but this time with the phase angle altered to produce the optical potential $\mathcal{V}^-$. Note that in the standard quantum walk procedure the wave-function is localized inside even positions for even steps while it sits in odd positions for odd steps of the walk. Hence the two effective translation operators $\hat{T}^+$ and $\hat{T}^-$, corresponding to optical potentials $\mathcal{V}^+$ and $\mathcal{V}^-$, should be applied in succession to implement odd and even steps of the walk respectively. Figure 3.72 represents the simulated probability distribution of the atomic wave-function $\psi_{10}$ after ten steps, where the authors use $J_{\max}/J_{\min} \approx 0.001$, $\mathcal{V}_{\max} = 25E_R$ and $\mathcal{V}_{\min} \approx 17.5E_R$.

A practical implementation of this scheme requires adiabatic modulation from an optical lattice to a super-lattice in order to avoid heating the trapped atom. The authors proposed an experimentally achievable trapping frequency of $\omega_T = 30\,\text{kHz}$ for which a suitable time of $\approx 33\,\mu\text{s}$ for adiabatic evolution can be employed. Other sources of error involved in this implementation include the instability of laser beams (e.g. uncontrollable phase shifts) which can influence the mobility of trapped atoms, causing imperfect manipulations during the walk, and incomplete tunneling produced due to the imperfect modulation of trapping potentials. The authors model the latter, depicted in Fig. 3.72, by using a tunneling time $t = J_{\min} + \delta t$, where $\delta t$ is the error in the modulation time of the beams.

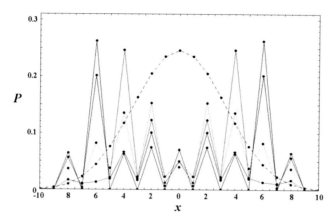

**Fig. 3.72** Probability distributions for $N = 10$ steps of the walk without errors $\delta t_0 = 0$ (*black*) and with errors $\delta t_0 = 0.2/J_{\min}$ (*blue*), $0.4/J_{\min}$ (*green*), and $0.6/J_{\min}$ (*red*). The *dashed line* shows the Gaussian probability distribution for $N = 10$ in a classical walk (From Joo et al. (2007))

A very different implementation scheme for a *continuous-time* quantum walk was proposed by Côté et al. (2006) based on ultra cold Rydberg (highly excited) $^{87}$Rb atoms in an optical lattice. These atoms have exaggerated properties. Taking $n$ to be the principal quantum number for the energy level where the excited outer electron resides, their radius and dipole moment $\mu$ scale as $n^2$, while their long lifetime $\tau_R$ scales as $n^3$. At large inter-atomic separations $R$, the interaction energy between two identical atoms depends on their states: for $\Delta n \equiv |n - n'| \simeq 0$ this interaction is very strong but becomes rapidly small as $\Delta n$ grows and the wavefunction overlap becomes negligible. As a result the interaction between two atoms which are both in a Rydberg state is exceptionally strong, spanning over many tens of microns while the interaction between Rydberg and ground-state atoms is much weaker, since $\Delta n$ is large.

The strong interaction between Rydberg atoms leads to the so-called blockade mechanism where, in an ensemble containing many atoms and under the right conditions, no more than one atom can be excited into a Rydberg state. Côté et al. (2006) identified two types of blockade, namely the dipole blockade and the van der Waals blockade. The principle underlying both mechanisms however is the same; strong Rydberg-Rydberg interactions shift the energy levels. So one atom can be resonantly excited into a Rydberg state, but additional Rydberg excitations of nearby atoms are prevented by the large shifts as depicted in Fig. 3.73. The van der Waals blockade is short-range and effective in a limited volume depending on $n$ and the laser bandwidth $\gamma_L$. For $n = 80$ and $\gamma_L = 1\,\mathrm{MHz}$, a conservative estimate gives a blockaded volume $4.6\,\mu\mathrm{m}$ in diameter, sufficient to span a single site of a $CO_2$ optical lattice. The dipole blockade has longer range interactions but relies on a frequency tuned external electric field and can therefore be turned on and off.

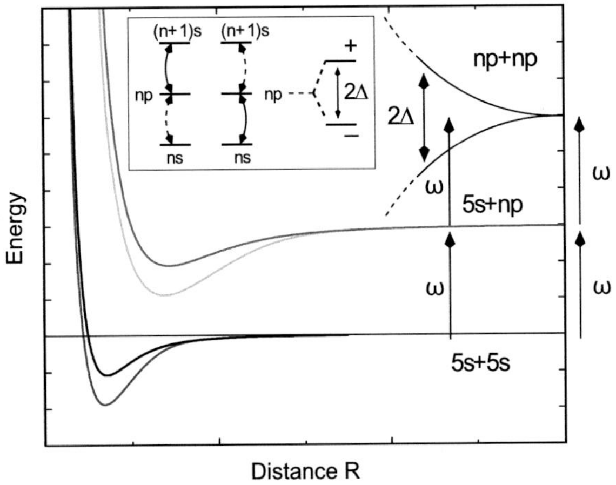

**Fig. 3.73** Sketch of blockade mechanisms: Rydberg-Rydberg interactions shift energy levels and prevent excitation of more than one atom. The *inset* shows a dipole blockade; on resonance, two degenerate $|np\rangle$ states split into $|+\rangle$ and $|-\rangle$ (From Côté et al. (2006))

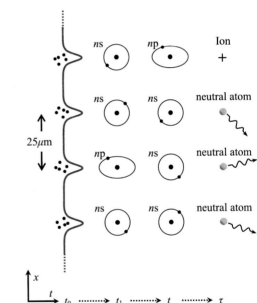

**Fig. 3.74** Scheme for implementing a quantum walk with Rydberg atoms (Adapted from Côté et al. (2006))

Figure 3.74 illustrates the principle behind the proposal of Côté et al. (2006) for implementing the continuous-time quantum walk. First, $N$ mesoscopic ensembles of $^{87}$Rb Rydberg atoms are localized by an optical lattice. Consequently in $N-1$ ensembles, one atom is promoted to a Rydberg state $ns$. In the remaining ensemble, one atom is pumped into $np$. Due to the strong interactions between the Rydberg atoms, the $np$ excitation will rapidly transfer from site to site by resonant transfer, $n_A s + n_B p \longrightarrow n_A p + n_B s$. The diffusion of the $np$ excitation among the sites throughout the lattice models the continuous-time quantum walk. The timescale for this excitation diffusion process is given by $\tau_{\text{hop}} \sim h/4V_{\text{dip}}$, where $V_{\text{dip}} \sim n^4/R^3$

**Fig. 3.75** Energy levels for
the excitation sequence
(Adapted from Côté et al.
(2006))

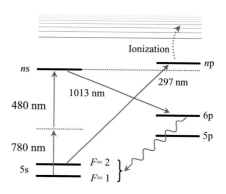

and $R$ is the distance between the neighboring ensembles. After a period such that many steps have occurred, the location of the $n$p state is measured. By repeating this sequence, one can determine the quantum walk distribution function.

The authors proposed loading the cold atoms inside a standard $CO_2$ optical lattice for which the trapping sites are separated by $5.3\,\mu$m. Initially there is an average of 5 atoms per site and all the atoms are in the $F = 2$ hyperfine level of the ground state 5s. Atoms are then eliminated except in every fifth site (spacing $26.5\,\mu$m), in order to achieve a better fractional definition of the atom separation, a more practical timescale for the steps in the quantum walk ($\tau_{hop} \sim 170$ ns for $n = 70$), and an easier site-selective state preparation and detection. This can be accomplished with a mask (or holographic phase plate) imaged onto the lattice and then illuminating the selected sites to transfer the atoms into $F = 1$. Atoms in all other sites could be 'blown away' using the cycling transition ($F = 2 \longrightarrow F' = 3$). From the regular array of ensembles in every fifth site a single ensemble, selected to be the origin of the walk, is transferred back into $F = 2$ using another mask. The remaining $N - 1$ ensembles in state 5s $F = 1$ are promoted to the 70s Rydberg state, using transform-limited pulses of light at 780 and 480 nm (near-resonant with the 5s $\longrightarrow$ 5p and 5p $\longrightarrow$ 70s transitions, respectively) to drive the two-photon transition to 70s. Finally a light at 297 nm, generated by nonlinear mixing of the 780 and 480 nm pulses, drives the one-photon transition of the remaining ensemble in state 5s $F = 2$ to the 70p state. These transitions are depicted in Fig. 3.75. It is important to note that a key requirement for the lattice is that each of the $N$ sites has exactly one excited atom: one in 70p and $(N-1)$ in 70s. The van der Waals blockade mechanism will prevent multiple excitations at a given site. During the walk, the lattice trap must be turned off because this low-frequency radiation is actually an anti-trap for Rydberg atoms. However, the motion of the cold atoms is negligible on the timescale of the walk.

The detection process involves applying a laser light at 1,013 nm to selectively dump the 70s population to the $6p_{3/2}$ level, which rapidly decays (in $\sim$100 ns) to the 5s ground state. The 70p atom is then ionized by a ramped electric field and the ion is detected with a position sensitive micro-channel plate detector (quantum efficiency $\sim$50 %). Ion optics magnification by a factor of 10 will allow individual

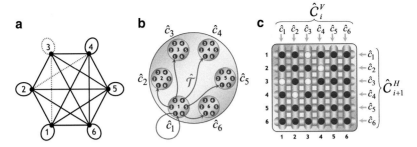

**Fig. 3.76** (a) A complete six-graph. Any generalized graph can be constructed by removing edges (*dotted lines*) from a corresponding complete graph. (**b**) Quantum walk Hilbert space $\mathcal{H}$, vertex coin operators $\hat{c}_1 \ldots \hat{c}_{\mathcal{N}}$ and the translation operator $\hat{T}$ implemented in the form of a particular mapping $\hat{T}|j,k\rangle \longrightarrow |k,j\rangle$ (Watrous 2001; Kendon 2006a). (**c**) The operator $\hat{T}$ is replaced by alternating the direction in which the global coin operator $\hat{C}$ is applied in successive steps of the walk. *Dotted circles* represent the states corresponding to edges removed from the complete graph. Instead of removing these unwanted states from $\mathcal{H}$, they are effectively barred from interacting with other states by appropriately designing the coin operators (From Manouchehri and Wang (2009))

sites to be resolved. An important aspect of the detection scheme in this proposal is that imperfect runs can be rejected. The authors account for two such scenarios:

- In some cases the blockade mechanism may not work perfectly leading to zero or multiple 70p excitations. These can be detected as zero or multiple ionizations and the run is rejected. Temporal resolution will allow free ions due to auto-ionization to be identified.

- While the walk is taking place, there may be ground state atoms left behind after the Rydberg excitation. To detect this scenario an imaging technique is utilized. These ground state atoms are removed prior to the detection process by an intense standing wave at 780 nm which will rapidly transfer enough momentum to these nonparticipating atoms to cause them to leave their sites. After the 70s atoms have returned to the ground state and the lattice is turned back on, a near-resonant light (with repumping) illuminates the entire sample and the resulting resonance fluorescence will allow imaging of the individual lattice sites. Since the walker will have been ionized, its final location will show up as a vacancy. But there would also be other vacancies detected if there were any nonparticipating ground state atoms which would have been removed from their sites by the 780 nm laser (the imaging should be delayed to allow their exit). Hence only those runs where a single ion is detected and a single vacancy is imaged in the lattice will be considered legitimate.

More recently Manouchehri and Wang (2009) have described a universal framework for performing a coined quantum walk on arbitrarily complex undirected graphs. A key feature of their proposal is to represent the Hilbert space of the walk as a two-dimensional $\mathcal{N} \times \mathcal{N}$ grid (see Fig. 3.76) where each individual node on the graph corresponds to a grid row (column) and the coin states within that node

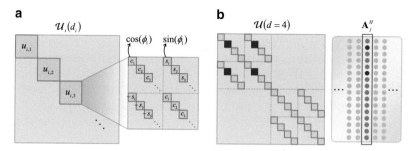

**Fig. 3.77** (a) The structure of a general matrix $\mathcal{U}_i(d_i > 2)$ resulting from the CS decomposition. (b) Structure of the matrix $\mathcal{U}_i(d_i > 2)$ acting on vector $\mathbf{A}_j^H$ for $d_i = 4$ (From Manouchehri and Wang (2009))

are the individual grid elements along that row (column). Manouchehri and Wang (2009) also showed that the standard quantum walk evolution given by Eq. 1.24 can be effectively reduced to

$$|\psi_n\rangle = \hat{C}_n^V \hat{C}_{n-1}^H \ldots \hat{C}_2^V \ \hat{C}_1^H \ |\psi_0\rangle, \tag{3.149}$$

where $\hat{C}_i^H$ ($\hat{C}_i^V$) correspond to the application of the coin operator for the $i$th step, along the horizontal rows (vertical columns) of the grid. A physical realization of such a quantum walk relies on the ability to implement the action of an arbitrary $\mathcal{N}$-level unitary operator $\hat{c}_j^H$ ($\hat{c}_j^V$) on the $\mathcal{N}$ nodes along the $j$th row (column) of the grid. Manouchehri and Wang (2009) showed that this can be achieved via a CS decomposition (Sutton 2009) which essentially reduces a general $\mathcal{N}$-level rotation matrix to a series of pair-wise or qubit rotations which can in principle be readily implemented. What makes this implementation scheme non-trivial however is the fact that the resulting pair-wise interactions are not limited to neighboring nodes. More precisely it can be shown that for a general rotation of column $j$ for example, we have

$$\hat{c}_j^v = \prod_{i=1}^{\mathcal{N}-1} \mathcal{U}_i(d_i), \tag{3.150}$$

for $d_i \in [2, 4, \ldots \mathcal{N}/2]$, where the action of each $\mathcal{U}_i(d_i)$ on the nodes of row $j$ consists of $\mathcal{N}/2$ simultaneous pair-wise interactions between nodes $kd_i + r$ and $kd_i + r + d_i/2$ for $k = 0, \ldots, \mathcal{N}/d_i - 1$ and $r = 1, 2 \cdots d_i/2$. Clearly for all $d_i \neq 2$, interactions are non-neighboring but follow a systematic pattern, depicted in Fig. 3.77.

On the basis of this universal quantum walk architecture, Manouchehri and Wang (2009) provided a supporting physical implementation which utilizes a BEC trapped in a 2D optical lattice. In developing this implementation scheme, the authors described a key process for applying a unitary rotation to a BEC initially prepared in the internal state $|0\rangle \equiv |F = 1, m_F = 1\rangle$ and distributed between two lattice sites $|j, k\rangle$ and $|j, k'\rangle$ such that $|\psi_0\rangle = \alpha_k |j, k\rangle \otimes |0\rangle + \alpha_{k'} |j, k'\rangle \otimes |0\rangle$. Such BEC localization can be accomplished using scanning electron microscopy (Würtz et al. 2009; Gericke et al. 2008) as illustrated in Fig. 3.78. The amplitudes $\alpha_k$ and $\alpha_{k'}$ can now be manipulated according to any desired unitary transformation in five steps depicted in Fig. 3.79. (1) Using a three-photon STIRAP (see Appendix C.2.3) to apply a $\pi$-rotation to the BEC at $|j, k\rangle$ which transfers it entirely to the internal state $|1\rangle \equiv |F = 2, m_F = 2\rangle$ and the new state of the system becomes $|\psi_1\rangle = \alpha_k |j, k\rangle \otimes |1\rangle + \alpha_{k'} |j, k'\rangle \otimes |0\rangle$. (2) Making use of spin(state)-dependant transport (see Appendix E.3) to fully overlap the two wave-packets at $|j, k'\rangle$ (selected as the stationary reference frame) and hence $|\psi_2\rangle = |j, k'\rangle \otimes (\alpha_k |1\rangle + \alpha_{k'} |0\rangle)$. (3) Using another three-photon STIRAP to perform the desired unitary rotation such that $|\psi_3\rangle = |j, k'\rangle \otimes (\tilde{\alpha}_k |1\rangle + \tilde{\alpha}_{k'} |0\rangle)$. (4) Reversing the spin(state)-dependant transport to bring the new BEC amplitudes $\tilde{\alpha}_k$ and $\tilde{\alpha}_{k'}$ back to their original sites, i.e. $|\psi_4\rangle = \tilde{\alpha}_k |j, k\rangle \otimes |1\rangle + \tilde{\alpha}_{k'} |j, k'\rangle \otimes |0\rangle$. (5) Finally performing another $\pi$-rotation on the state $|j, k\rangle$ to transfer the BEC back to the internal state $|0\rangle$ producing the desired outcome $|\psi_5\rangle = \tilde{\alpha}_k |j, k\rangle \otimes |0\rangle + \tilde{\alpha}_{k'} |j, k'\rangle \otimes |0\rangle$. This scheme can be readily extended to simultaneously activate all the pair-wise interactions required for performing the global coin operations. The authors also point out that, although the control laser wavelength and the lattice period $\lambda_{\text{lattice}}$ are comparable in size, problems associated with unwanted interactions of the control laser with neighboring sites can be circumvented by adopting techniques such as those detailed in Cho (2007) and Gorshkov et al. (2008) or more readily by choosing every 2nd, 3rd or $\ell$th lattice site to represent the walk states.

At the conclusion of the walk, BEC densities throughout the lattice can be determined via scanning electron microscopy (Würtz et al. 2009; Gericke et al. 2008) or spin-selective absorption imaging (Greiner et al. 2003), although the latter requires repeated runs of the experiment for each node density measurement. The corresponding quantum walk distribution is then derived by integrating the BEC amplitudes over an area $\ell \lambda_{\text{lattice}} \times \ell \lambda_{\text{lattice}}$ centered around the key lattice sites.

Finally, Manouchehri and Wang (2009) showed that the proposed quantum walk scheme offers a polynomial speedup over an equivalent quantum circuit implementation, highlighting the expected trade off between resource and time scalability. They also highlighted the BEC internal state phase decoherence times that are presently in the order of a few ms (Lee et al. 2007) as compared with a single-site transport time of $\sim$50 $\mu$s (Mandel et al. 2003a,b) and STIRAP pulse durations of $\sim$60 $\mu$s (Wright et al. 2008). This together with the successful realization of spin(state)-dependant BEC transport for up to seven sites reported in Mandel et al. (2003a), led the authors to conclude that a "proof of principle" implementation (i.e. the first few steps of the walk on an arbitrary graph with a few nodes) should indeed be possible, utilizing the existing experimental techniques.

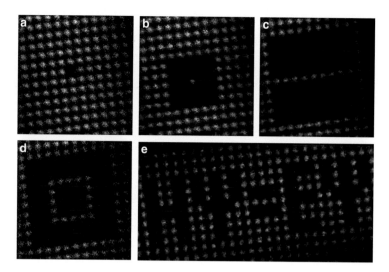

**Fig. 3.78** Patterning a Bose-Einstein condensate in a 2D optical lattice with a spacing of 600 nm. Every emptied site was illuminated with the electron beam (7 nA beam current, 100 nm FWHM beam diameter) for 3 ms (**a** and **b**), 2 ms (**c** and **d**), and 1.5 ms (**e**), respectively. The imaging time was 45 ms. Between 150 and 250 individual images have been summed for each pattern (From Würtz et al. (2009))

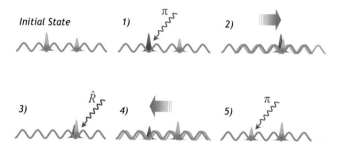

**Fig. 3.79** Steps for applying a unitary transformation to BEC amplitudes trapped in a pair of non-neighboring optical lattice sites (From Manouchehri and Wang (2009))

## 3.7 Solid State

In a related work, Manouchehri and Wang (2008b) proposed another physical implementation based on the universal framework introduced in Manouchehri and Wang (2009), this time utilizing a 2D array of interacting quantum dots. In this scheme the initial state of the walk is represented by the distribution of a single electron wave-function throughout a $2\mathcal{N} \times 2\mathcal{N}$ quantum dot grid, where every second row (column) of dots is used as a temporary "register". Manouchehri and Wang (2008b) described two mechanisms: (1) how to implement a pair-wise

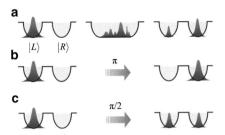

**Fig. 3.80** (a) Arbitrary qubit rotations can be performed by a precision control of the central potential barrier. (b) The barrier raised in time to implement a $\pi$ rotation (also known as a SWAP or NOT gate) in which the electron wave-function is transferred entirely from one dot to its neighbouring dot. (c) The barrier is raised in time to implement a $\pi/2$ rotation resulting in a 50–50 split of the wave-function initially confined to either of the dots

interaction between any two neighboring quantum dots, and (2) how to extend this to non-neighboring dots.

A pair of neighboring quantum dots can be made to undergo a unitary rotation $\hat{R}$ by the precision control of the potential barrier between them (Green and Wang 2005; Dinneen and Wang 2008). Assuming that the electron wave-function is initially confined to the left dot, i.e. $|\psi_0\rangle = |L\rangle$, by dropping the barrier it is free to move between the two dots as depicted in Fig. 3.80. Returning the barrier to its initial state at a precise moment in time makes it possible to recapture the wave-function but this time with the desired distribution across the two dots given by

$$|\psi_T\rangle = \alpha|L\rangle + \beta|R\rangle, \qquad (3.151)$$

where $\alpha$ and $\beta$ are complex amplitudes. This scheme can be further enhanced by carefully positioning a control charge near the central barrier. Manouchehri and Wang (2008b) carried out a numerical simulation of this process by solving the many-electron Schrödinger equation

$$i\hbar \frac{\partial}{\partial t} \psi\left(\vec{r}_1, \vec{r}_2, \ldots t\right) =$$

$$\left[ \sum_{i=1}^{N} \left( -\frac{\hbar^2}{2m^*} \nabla_{r_i}^2 + V(\vec{r}_i) \right) + \frac{e^2}{4\pi\epsilon} \sum_{i>j=1}^{N} \frac{1}{r_{ij}} \right] \psi\left(\vec{r}_1, \vec{r}_2, \ldots t\right), \quad (3.152)$$

for $N = 2$ distinguishable electrons, where $\vec{r}_1$ and $\vec{r}_2$ correspond to the coordinates of the target and control electrons, $V(r_i)$ is the confinement potential due to an external field experienced by each electron, $r_{12} = |\vec{r}_2 - \vec{r}_1|$, and the last term $\frac{e^2}{4\pi\epsilon r_{12}}$ represents the electron-electron interaction. In this model all electrons in the system are treated on equal footing, and evolve coherently in time under the influence of each other as well as the external field. An efficient numerical solution to the above equation can be found using a Chebychev polynomial expansion (Wang and Midgley 1999; Green and Wang 2005)

$$\psi\left(\vec{r}, t\right) = \exp\left(-i\left(E_{\max} + E_{\min}\right)t/2\right) \sum J_n(\alpha) T_n(-i\tilde{H}) \psi\left(\vec{r}, 0\right), \qquad (3.153)$$

where $E_{\max}$ and $E_{\min}$ are the upper and lower bounds on the energies sampled by the numerical grid, $J_n(\alpha)$ are Bessel functions of the first kind, and $T_n$ are the Chebyshev polynomials. The normalized Hamiltonian is defined as

$$\tilde{H} = \frac{2H - E_{\max} - E_{\min}}{E_{\max} - E_{\min}}, \qquad (3.154)$$

to ensure convergence. Figure 3.81 presents an example of the resulting time evolution in real space for an electron confined in a coupled quantum dot system. A wide range of qubit rotations in the Bloch sphere (see Appendix C) can be performed by controlling the central potential barrier and the electron-electron interactions, as demonstrated in Fig. 3.82.

To interact a pair of *non-neighboring* quantum dots Manouchehri and Wang (2008b) proposed moving the two dots close to each other via the temporary register, then applying the desired rotation to the now-adjacent dots, and finally returning them to their original location. What makes this process systematic and practically viable is the specific pattern of pairwise interactions arising from Eq. 3.150. As depicted in Fig. 3.83, the action of each $\mathcal{U}_i(d)$ can be implemented in five steps:

1. Apply $\mathcal{N}/2$ simultaneous $\pi$ rotations to pairs of quantum dots at $kd + r$ and their adjacent register dot. This has the effect of transferring the electron wave-packets entirely to their neighboring register dots.
2. Adiabatically move the register quantum dots, much like a "conveyor belt" carrying the electron wave-packets along the register row (column). This can be achieved by carefully designing and manipulating the voltage applied to the electrodes such that the confinement potentials experience an effective motion. Moving the register dots over a distance equal to $d$ times the quantum dot width would effectively allow the amplitudes at $kd + r$ to be coupled with the amplitudes at $kd + r + d/2$.
3. Simultaneously apply $\mathcal{N}/2$ general rotations $\hat{R}$ to the recently coupled quantum dot pairs by controlling the barrier between them.
4. Return the register qubits to their original location by reversing the adiabatic motion.
5. Introduce another $\pi$ rotation to move the amplitudes from register dots back to their original positions.

It should be emphasized that the above steps can be carried out simultaneously throughout the entire grid and Bloch rotations $\hat{R}$ in step 3 can be different for each quantum dot pair.

Utilizing these quantum dot interactions it is possible to manipulate the electronic wave-function in a manner which corresponds exactly to the quantum walk

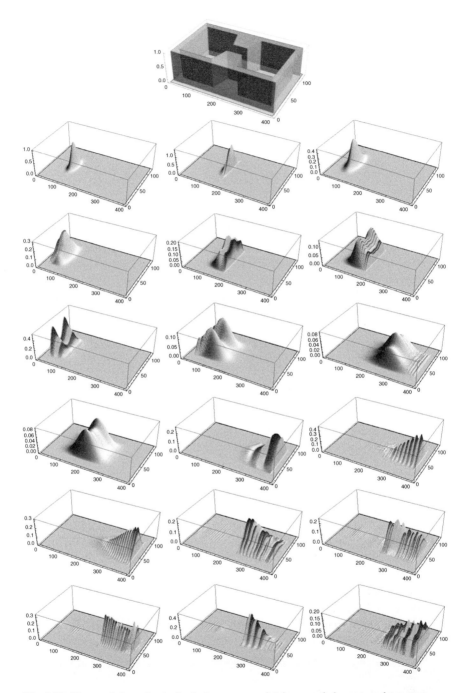

**Fig. 3.81** Time evolution of a single electron wave-packet in a coupled quantum dot system

**Fig. 3.82** Qubit rotations in the Bloch sphere as a function of time. Time evolution of amplitudes $\alpha$ and $\beta$ in Eq. 3.151 as well as the phase angle between them (From Green and Wang (2005))

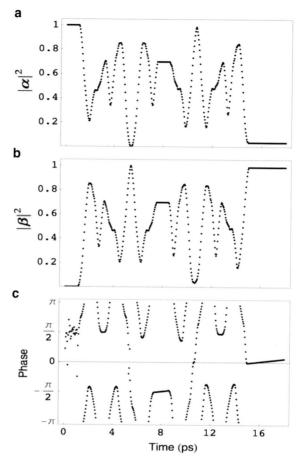

evolution described in Eq. 3.149 (Manouchehri and Wang 2009). After repeated applications of the above five steps, a measurement of the charge distribution throughout the grid corresponds to the final quantum walk probability distribution.

Earlier, Manouchehri and Wang (2008a) proposed another design for implementing a coined quantum walk on a line using a 1D array of quantum dots. In this scheme, illustrated in Fig. 3.84, the quantum walk is represented by the propagation of a single electron wave-function through the array of dots using a series of specially optimized two- and three-photon $\Lambda$ STIRAP (see Appendix C.2.3) responsible for performing the translation and coin operations respectively.

Initially all odd barriers have been significantly lowered creating pairs of coupled dots that form the nodes of the quantum walk. An essential feature of this design is the dimensions of the quantum dots as well as the potential barrier energies which are specifically engineered in such a way that for each node $i$ lower energy states labeled by $|\uparrow, i\rangle$, $|\downarrow, i\rangle$ and $|A, i\rangle$ are, to a very good approximation, spatially separable and the electron wave-function is localized within the dot. For

**Fig. 3.83** The initial electron wave-function inside the quantum dot grid followed by five steps required to implement the operator $\mathcal{U}_i(d = 4)$

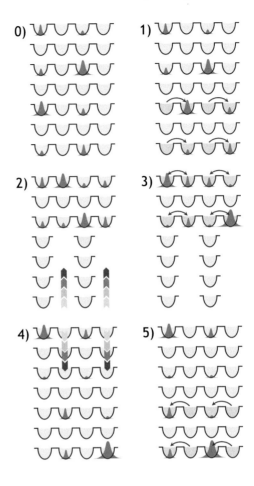

energies above their joint potential barrier however the two dots share common electronic states including an excited state $|e, i\rangle$. Another significant consideration in the design of the quantum dots is the ability to perform selective addressing of states which are being coupled via STIRAP and to avoid all unwanted secondary excitations. Manouchehri and Wang (2008a) described how this may be achieved experimentally using simple quantum dot structures.

Each step of the walk is performed in five stages.

1. A three-photon STIRAP operation performs an arbitrary coin rotation $\hat{C}$ (Kis and Renzon 2002), simultaneously on all coin states $|\uparrow, i\rangle$ and $|\downarrow, i\rangle$ facilitated by the auxiliary state $|A, i\rangle$.
2. A two-photon STIRAP will then transfer all the $|\uparrow, i\rangle$ states to their corresponding $|A, i\rangle$ state.
3. All the odd barriers are then adiabatically raised while the even barriers are lowered, virtually reversing the pairing of the quantum dots. In this new

**Fig. 3.84** The sequence of
two- and three-photon
STIRAP to implement a
single step of the quantum
walk. (**a**) The initial state of
the walk with the electron
confined to state $|\uparrow, 1\rangle$.
(**b**) A three-photon STIRAP
implements the coin rotation
$\hat{C}$, mixing the states $|\uparrow, i\rangle$
and $|\downarrow, i\rangle$. (**c**) A two-photon
STIRAP transfers the
population from state $|\uparrow, i\rangle$
to state $|A, i\rangle$. (**d**) Even
barriers are lowered and odd
barriers are raised in order to
regroup the quantum dots. It
is now possible for another
two-photon STIRAP to
transfers the population from
state $|A, i\rangle$ to state
$|\uparrow, i+1\rangle$, completing the
one-sided translation
operation $\hat{T}^{\uparrow}$. (**e**) Potential
barriers are returned to their
initial setting and the above
process repeated (From
Manouchehri and Wang
(2008a))

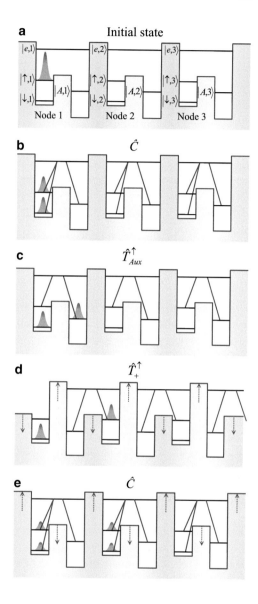

arrangement every state $|A, i\rangle$ previously associated with the $i$th node is now
paired up with $(i + 1)$th node.
4. Using a second two-photon STIRAP the amplitude in state $|A, i\rangle$ is then
   transferred to $|\uparrow, i+1\rangle$ which completes the implementation of a one-sided
   translation operator $\hat{T}^{\uparrow}$. The authors showed that the resulting quantum evolution
   is in fact identical to that of a standard quantum walk, up to a translation and
   relabeling of nodes.

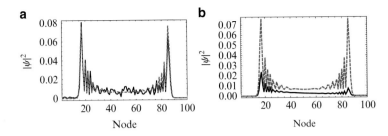

**Fig. 3.85** Deviation from the exact quantum walk distribution after 100 steps (*dashed*) due to the introduction of (**a**) 5 % uncertainty in the laser pulse phases (Manouchehri and Wang 2008a) and (**b**) radiative decay $\hbar\Gamma_{Aux} = 4 \times 10^{-4}$ meV out of the auxiliary states

5. Finally the barriers are returned to their original setting and the process repeated for additional steps. The final quantum walk distribution corresponds to the probability distribution for detecting the electron inside each node in the array of dots.

Manouchehri and Wang (2008a) also simulated the effect of noise disturbance and experimental uncertainty on the resulting distribution and demonstrated a relatively robust response against the introduction of white noise in the laser pulse parameters, as evident in Fig. 3.85a for example. Radiative decay out of various quantum states involved in the STIRAP operations can also impact on the quantum walk's fidelity. Although the process is, to a large extent, unaffected by a leaky excited state $|e, i\rangle$ (see Appendix B.3.3), modifying the STIRAP Hamiltonian to include a decay term corresponding to the auxiliary state $|A, i\rangle$ allows for a straightforward simulation of this process as depicted in Fig. 3.85b. Noticeably, in this case while the distribution is clearly attenuating, the characteristic quantum walk signature remains intact. The distribution is asymmetric owing to the additional leaky shift operations required to move the walker along the nodes from left to right.

Solenov and Fedichkin (2006a) proposed using a ring shaped array of identical tunnel-coupled quantum dots to implement the *continuous-time* quantum walk on a circle. Their scheme is loosely based on the work of de la Torre et al. (2003) which considered the quantum diffusion of a particle in an initially localized state, on a cyclic lattice with $N$ sites. In Solenov and Fedichkin's proposal (2006a), the walk is performed by an electron initially placed in one of the dots. Each dot is continuously monitored by an individual point contact which, as discussed by Gurvitz et al. (2003), introduces some decoherence in the electron's evolution. The authors point out that the quantum dot cycle with "attached" point contacts can, in principle, be fabricated with the help of gate-engineering techniques in semiconductor heterostructures, allowing the formation of quantum dots and point contacts electrostatically, by placing metal gates on the structure with a 2D electron gas. By changing the potential on the gates one can allocate areas of 2D electron gas, creating the necessary confinement profile. The simplest example of such a

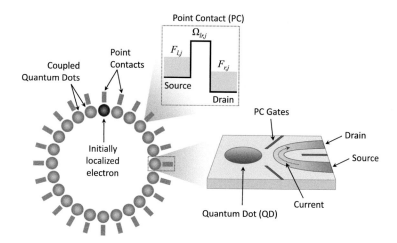

**Fig. 3.86** Continuous-time quantum walk architecture: ring of quantum dots, each of which is monitored by the corresponding point contact that introduces decoherence. $F_{l,j}$ and $F_{r,j}$ are chemical potentials of the source and drain of the $j$th point contact. The presence of an electron in the $j$th quantum dot affects the source-to-drain tunneling amplitude $\Omega_{lr,j} \longrightarrow \Omega_{lr,j} + \delta\Omega_{lr,j}$ of the $j$th point contact (Adapted from Solenov and Fedichkin (2006a))

structure containing two quantum dots was investigated experimentally in Pioro-Ladrière et al. (2005). The key assumptions in this scheme are that (1) identical point contacts are formed far enough from the quantum dot structure so that the tunneling between them is negligible and (2) Coulomb interaction between electrons in the quantum dot and point contact is taken into account.

The quantum evolution of this system involves three constituent parts, the quantum dots, the point contacts and the interaction between the two. The Hamiltonian of an electron placed in the quantum dot cycle is

$$\hat{H}_{\text{cycle}} = \frac{1}{4} \sum_{j=0}^{N-1} (\hat{a}_{j+1}^\dagger \hat{a}_j + \hat{a}_j^\dagger \hat{a}_{j+1}), \qquad (3.155)$$

where $\hat{a}_j^\dagger$ and $\hat{a}_j$ are the creation and annihilation operators for an electron on site $j$, $N$ is the number of quantum dots in the cycle, and $\hat{a}_N \equiv \hat{a}_0$. The point contact, placed next to each quantum dot, consists of two reservoirs of electrons, source and drain, that are coupled through the potential barrier shaped by the gates as depicted in Fig. 3.86. The Hamiltonian of the $j$th point contact can be written as

$$\hat{H}_{\text{PC},j} = \sum_l E_{l,j} (\hat{a}_{l,j}^\dagger \hat{a}_{l,j}) + \sum_r E_{r,j} (\hat{a}_{r,j}^\dagger \hat{a}_{r,j}) + \\ \sum_{lr} \Omega_{lr,j} (\hat{a}_{l,j}^\dagger \hat{a}_{r,j} \hat{a}_{r,j}^\dagger \hat{a}_{l,j}), \qquad (3.156)$$

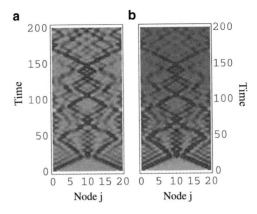

**Fig. 3.87** Probability distribution along the cycle as a function of time and node number, for $N = 20$ and a decoherence factor $\Gamma = 0$ (**a**), and 0.01 (**b**). Here $j \in [0, N-1]$ stands for the node number and *darker regions* denote higher probabilities. The electron is initially placed at $j = 0$. The probability distribution of the walk with some decoherence added, (**b**), converges to a uniform distribution, i.e., to $1/N$ (From Solenov and Fedichkin (2006a))

where $\hat{a}^{\dagger}_{l,j}$, $\hat{a}_{l,j}$, $\hat{a}^{\dagger}_{r,j}$ and $\hat{a}_{r,j}$ are creation and annihilation operators in the left (source) and right (drain) reservoirs of the $j$th point contact, $\Omega_{lr,j}$ are the tunneling amplitudes between states $l$ and $r$ of the $j$th point contact and the interaction energies $E_{l,j}$ and $E_{r,j}$ between the source and drain reservoirs are determined by chemical potentials $F_{l,j}$ and $F_{r,j}$. Finally a weak Coulomb interaction between electrons in the point contact and quantum dot, makes it possible to observe the presence of the electron in the $j$th quantum dot as it raises the tunneling amplitude $\Omega_{lr,j}$ by an amount $\delta\Omega_{lr,j}$ which is assumed to be small compared to the other amplitudes in the problem. This process, which can be represented via the Hamiltonian

$$\hat{H}_{\text{int},j} = \sum_{lr} \delta\Omega_{lr,j}\hat{a}^{\dagger}_{j}\hat{a}_{j}(\hat{a}^{\dagger}_{l,j}\hat{a}_{r,j} + \hat{a}^{\dagger}_{r,j}\hat{a}_{l,j}), \qquad (3.157)$$

introduces weak measurement on the electron in each node of the graph, and, therefore, results in some loss of coherence in electron evolution. The total Hamiltonian for the system is therefore given by

$$\hat{H} = \hat{H}_{\text{cycle}} + \sum_{j=0}^{N-1}(\hat{H}_{\text{PC},j} + \hat{H}_{\text{int},j}). \qquad (3.158)$$

The authors employed the reduced density matrix formalism in conjunction with Bloch-type rate equations to obtain the probability distribution $P_j(t)$, representing the probability for an electron, initially placed at node 0, to be found on node $j$ at time $t$. The probability distribution is shown in Fig. 3.87. Figure 3.87a shows the evolution of the walk in the absence of decoherence. Figure 3.87b shows the evolution when the system is exposed to weak measurement (decoherence). It can be seen that a weak measurement of the system leaves the time evolution of the walk almost unchanged from that of the coherent walk. However the coherent oscillation pattern is gradually suppressed by effective averaging that leads to the onset of a uniform distribution.


## 3.8   Quantum Circuits

Košík and Bužek (2005) noted that building a whole network of multiports (Sect. 3.1.6) would need exponentially growing resources; e.g. for a $d$-cube the number of vertices $N = 2^d$ which clearly grows exponentially. However, to encode the states under consideration, we need only $d\lceil \log d \rceil$ qubits (d qubits for the position register and at least $\lceil \log d \rceil$ qubits for the direction register) and the authors described a network of quantum gates operating on the qubit register of this size to implement the walk. To do so a notation used by Shenvi et al. (2003) is employed where each node is labeled by a $d$-bit binary string. For a 4-cube for example, the nodes are labeled as 0000, 1000,...1110, 1111. Two nodes on the hypercube described by bit strings $\vec{x}$ and $\vec{y}$ are connected by an edge if $|\vec{x} - \vec{y}| = 1$, where $|\vec{x}|$ is the hamming weight of $\vec{x}$. In other words, if $\vec{x}$ and $\vec{y}$ differ by only a single-bit flip, then the two corresponding nodes on the graph are connected. Each of the $2^d$ nodes on the $d$-cube has degree $d$ (i.e. connected to $d$ other nodes). In this way the quantum walk can be described using the basis states $|\vec{x}, i\rangle$, where $\vec{x}$ is a bit string which specifies the position state on the hypercube, and $i = 1, 2, \ldots d$ specifies the direction of the walk by denoting the bit which must be flipped to represent a neighboring node. Taking the example of a 4-cube again, suppose we are at node $\vec{x} = 0000$. Using this notation we have

$$|\vec{x}, 1\rangle \equiv |0000, 1000\rangle \equiv |\vec{x}, \vec{y}_1\rangle, \tag{3.159}$$

$$|\vec{x}, 2\rangle \equiv |0000, 0100\rangle \equiv |\vec{x}, \vec{y}_2\rangle, \tag{3.160}$$

$$|\vec{x}, 3\rangle \equiv |0000, 0010\rangle \equiv |\vec{x}, \vec{y}_3\rangle, \tag{3.161}$$

$$|\vec{x}, 4\rangle \equiv |0000, 0001\rangle \equiv |\vec{x}, \vec{y}_4\rangle, \tag{3.162}$$

that is, all the states corresponding to the edges connecting node $x$ to its neighboring nodes $y_1$ through to $y_4$. We also note that if $|\vec{x}, i\rangle \equiv |\vec{x}, \vec{y}_i\rangle$, then $|\vec{y}_i, i\rangle \equiv |\vec{y}_i, \vec{x}\rangle$.

The authors then defined a position register $|S\rangle$ which uses $d$ qubits to encode the superposition of all the $d$-bit binary strings $|\vec{x}\rangle, |\vec{y}\rangle, \ldots$ representing the $2^d$ nodes and a direction register $|D\rangle$ which uses at least $\lceil \log d \rceil$ qubits to encode the superposition of basis states $|1\rangle, |2\rangle, \ldots, |d\rangle$. The action of quantum gates on the registers is given by a unitary operator $U$. The first part in applying $U$ is the controlled negation of each bit of $S$ depending on $|D\rangle$. The second part is the transformation of the state $|D\rangle$, so that the action of the multiports is accounted. More precisely, the first part is

$$|S, D\rangle \xrightarrow{\hat{S}} \sum_{i=1}^{d} |S'_i, i\rangle \langle i | D\rangle, \tag{3.163}$$

where $S'_i$ represent the neighboring nodes of $S$, and the second part reads

$$|S'_i, i\rangle \xrightarrow{\hat{M}} |S'_i\rangle \left[ r|i\rangle + t \sum_{\substack{j=1 \\ j\neq i}}^{d_x} |j\rangle \right], \tag{3.164}$$

for each $i = 1, 2 \ldots d_x$. Initializing the registers in Eq. 3.163 as $|S, D\rangle = |\vec{y}_i, i\rangle$ and letting $\vec{x}$ be the node connected to $\vec{y}_i$ via the $i$th bit flip, the equivalence of the above implementation with Eq. 3.67 becomes evident and we can formally identify $\hat{U}_x \equiv \hat{M}\hat{S}$. The first part described by Eq. 3.163 can be implemented using a variant of the controlled-NOT (CNOT) gate. The CNOT gate operates on two qubits such that it negates the first (target) qubit, if and only if (iff) the second (control) qubit is nonzero. The action of the CNOT gate is described by the two-qubit operator $C_{CNOT} = \sigma_x \otimes |1\rangle\langle 1| + \mathbb{1} \otimes |0\rangle\langle 0|$. Here the authors employed the $\phi_{CNOT}$ gate, which differs from the CNOT gate in that, unlike a single qubit, it has a $d$-dimensional control state. If the control state is $|\phi\rangle$ (the accepting control state), then the target qubit is negated, otherwise it is kept in the original state. The operational form of the $\phi_{CNOT}$ gate is

$$\phi_{CNOT} = \sigma_x \otimes |\phi\rangle\langle\phi| + \mathbb{1} \otimes (1 - |\phi\rangle\langle\phi|) \tag{3.165}$$

Equation 3.163 can then be implemented by using $d$ $\phi_{CNOT}$ gates depicted in Fig. 3.88. Each gate operates on a different qubit from the position register. If the gate operates on the $i$th qubit, the accepting control state is chosen to be $|i\rangle$.

Equation 3.164, on the other hand, is implemented by using a single unitary operator $M$ acting on the direction register. It corresponds to the transformation of the state due to the multiports and is equivalent to a $d \times d$ matrix

$$M = \begin{pmatrix} r & t & \cdots & t \\ t & r & & \\ \vdots & & \ddots & t \\ t & & t & r \end{pmatrix}. \tag{3.166}$$

Douglas and Wang (2009) developed a set of exact and efficient quantum circuits to implement discrete-time quantum walks on several highly symmetric graphs, such as a 16-length cycle (Fig. 3.89), a 2D hyper cycle, a twisted toroidal lattice

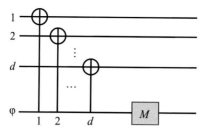

**Fig. 3.88** The gate which implements the scattering quantum walk on the $d$-dimensional hypercube. The input state is the position register ($d$ qubits labeled as $1, \ldots, d$) and the direction register $|D\rangle$. There are $d$ $\phi_{\text{CNOT}}$ gates stacked together, with accepting states $|1\rangle, \ldots, |d\rangle$, which change the position register, and the $M$ gate which changes the direction register (Adapted from Košík and Bužek (2005))

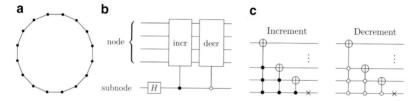

**Fig. 3.89** Quantum walk along a 16-length cycle (**a**) and its corresponding quantum circuit implementation (**b**). Quantum circuit implementations of the increment and decrement gates on $n$ qubits are shown in (**c**), producing cyclic permutations in the $2^n$ bit-string states (From Douglas and Wang (2009))

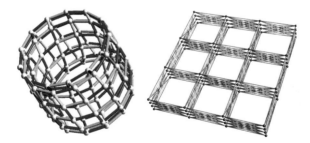

**Fig. 3.90** 'Twisted' toroidal lattice graph. Each node in the representation on the *left* contains four sub-nodes of the graph, as indicated on the *right* (From Douglas and Wang (2009))

graph (Figs. 3.90 and 3.91), a complete 16-graph, and a regular binary glued tree (Figs. 3.92 and 3.93), as well as the complete $3^n$ graphs (Fig. 3.94).

Loke and Wang (2011) extended this work to include non-regular graphs such as the star graphs of arbitrary degree and Cayley trees with an arbitrary number of layers. The examples considered in Douglas and Wang (2009) and Loke and Wang (2011) are quite simple, but more complex variations can still be efficiently implemented such as composites of highly symmetric graphs, symmetric graphs with a small number of 'imperfections', as well as graphs possessing a certain bounded level of complexity.

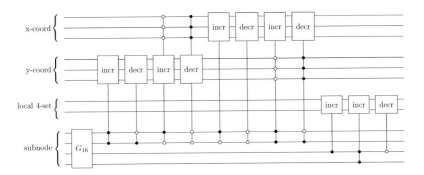

**Fig. 3.91** Quantum circuit implementing a quantum walk along the twisted toroidal of Fig. 3.90, with dimension $8 \times 8 \times 4$ (From Douglas and Wang (2009))

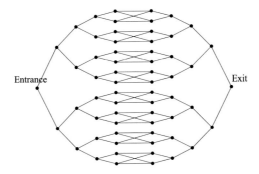

**Fig. 3.92** Binary glued trees with regular interconnections between the central levels (From Douglas and Wang (2009))

Quantum walks have been used to search for marked vertices along highly symmetric graphs, including the hypercube, complete graphs and complete multipartite graphs. These studies have dealt with the computational complexity of such searches relative to an oracle – looking at the number of steps of a quantum walk required to find a marked vertex, with individual steps of the walk itself largely left to the oracle. In such cases, searching using quantum walks has yielded a quadratic speedup over classical search algorithms.

In a practical implementation of such a search algorithm, the computations performed by the oracle (that is, performing a step of the walk in which the coin operator differs for marked and unmarked nodes) would of course affect the runtime. The work carried out by Douglas and Wang (2009) and Loke and Wang (2011) can be used to efficiently implement such an oracle – using $O(\log(n))$ elementary gates for a graph of order $n$ – given a highly symmetric graph such as those considered in Hines and Stamp (2007), Reitzner et al. (2009), Douglas and Wang (2009) and Loke and Wang (2011).

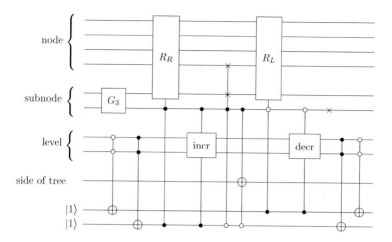

**Fig. 3.93** Quantum circuit implementing a quantum walk along a glued tree with a regular labeling of the nodes (From Douglas and Wang (2009))

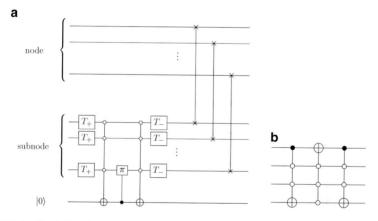

**Fig. 3.94** (**a**) Qutrit-based quantum circuit implementing a quantum walk along a complete $3^n$-graph. (**b**) A transposition of the $|0000\rangle$ and $|1001\rangle$ states (From Douglas and Wang (2009))

## 3.9   Recent Developments

In the preceding sections we have surveyed an exhaustive record of theoretical and experimental efforts aimed at implementing the quantum walk of a single particle. As we have seen, the exploration of this class of quantum walks, particularly those pertaining to graphs, has proven highly valuable in characterizing key features that may be harnessed for their potential algorithmic application. Moreover, while wide-ranging in their physical realization, approaches to implement a single-particle quantum walk in the laboratory have generated much insight into the requisite elements and pitfalls that can accompany attempts to build the necessary quantum walk hardware.

What is now emerging as a natural extension to the present body of work is an examination of quantum walks involving two or more correlated particles. Multi-particle quantum walks can offer potential advantages over the single-particle walk when considering the possibility of incorporating more exotic quantum behavior, such as entanglement as well as various forms of inter-particle interactions. These have been shown to greatly enrich the quantum walk's evolution and could in turn lead to a new class of applications (Berry and Wang 2011).

One of the earliest discussions of a two-particle quantum walk came from Pathak and Agarwal (2007) who considered a setup identical to the implementation of Do et al. (2005) in Fig. 3.21 which allowed a four dimensional coin space to be encoded using the photon direction as well as polarization states. The authors then analyzed the coined quantum walk of a pair of photons in a variety of initial states, including coherent as well as separable and entangled Fock states, and showed the remarkable dependence of the two photon detection probability on the quantum nature of the evolution. More specifically, Pathak and Agarwal (2007) demonstrated that while a pair of photons initially in separable coherent states remain separable throughout the walk, photon pairs that are in an inherently quantum state become entangled even if the states are initially separable, leading to joint probability distributions that no coherent states can reproduce. It should be emphasised here that although schemes for implementing two- or higher-dimensional quantum walks had been previously proposed by other authors including Roldán and Soriano (2005), those studies did not consider the critical impact of quantum entanglement on the quantum walk dynamics.

Peruzzo et al. (2010) experimentally demonstrated the continuous-time quantum walks of two identical photons in an array of coupled waveguides in a $SiO_xN_y$ chip (Fig. 3.95). Photon pairs are generated in a frequency entangled but spatial mode-separable state via type I spontaneous parametric downconversion. They are then injected into two of the three polarization-maintaining input fibers, butt-coupled to the waveguide chip in which the propagation constant $\beta$ and coupling constant $C$ between adjacent waveguides are designed to be uniform. At the output of the chip, an array of multimode fibers guides the photons to single-photon-counting modules used to postselect all possible two-photon coincidences between different outputs of the array. The measured correlation matrices (defined as the probability of detecting a two-photon coincidence across waveguides) for injecting two single photons into the waveguides $-1$ and $1$ are plotted in Fig. 3.96a for photons made distinguishable using temporal delay (not overlapped) and in Fig. 3.96c for pairs of indistinguishable (overlapped) photons. Respective correlations arising from ideal simulations are plotted in Figs. 3.96b and 3.96d. Peruzzo et al.'s demonstrations (2010) showed uniquely nonclassical behavior of two identical (indistinguishable) particles, tunneling through arrayed potential wells.

Rohde et al. (2011) presented a generalized framework for performing multi-particle coined quantum walks with an arbitrary number of walkers acting on

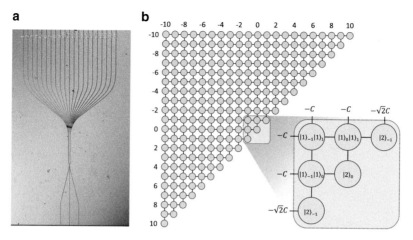

**Fig. 3.95** (**a**) An optical micrograph of a 21-waveguide array showing the 3 input waveguides, initially separated by 250 mm, bending into the 700-mm-long coupling region. All 21 outputs bend out to 125-mm spacing. (**b**) The two-dimensional lattice of vertices that represent the state space of 2 photons populating the 21 waveguides. (*Inset*) Enlarged portion of the lattice displays the vertex representation of the two photon basis states where the hopping amplitudes between adjacent vertices are either $-C$ or $-\sqrt{2}C$ as labeled (From Peruzzo et al. (2010))

arbitrary graph structures. The authors then outlined an illustrative example for implementing a two-photon quantum walk on a line using a network of beam splitters depicted in Fig. 3.97. In Sect. 3.1.4 we explored how similar setups have been employed to implement the quantum walk of a single particle on a line. Now supposing that the system contains exactly two indistinguishable photons, the beam splitter transformation can be described by an extended $6 \times 6$ matrix acting on basis states $|00\rangle, |01\rangle, |10\rangle, |11\rangle, |20\rangle$ and $|02\rangle$, where $|mn\rangle$ is the two-mode state with $m$ photons in the first mode and $n$ photons in the second. Crucially, a close examination of the beam splitter transformation matrix reveals that unlike the single photon case where the input-output relationship could be simulated using coherent light (Jeong et al. 2004), the two-photon transformation matrix gives rise to uniquely quantum interference effects which differentiate the single- and two-walker scenarios.

Such a differentiation is readily observable in Fig. 3.98 which draws a comparison between the joint probability distributions for a pair of distinguishable and indistinguishable photons participating in a four step quantum walk simulation. Whereas in the case of distinguishable walkers the coincidence matrix is uncorrelated and can be obtained by multiplying the single photon statistics along one axis by the single photon statistics along the other, in the case of indistinguishable walkers, the matrix is correlated as characterized by the suppression of the two diagonal peaks. The authors note that all the off-diagonal elements of the joint probability distributions can be experimentally obtained using coincidence measurements, which do not require number resolving detection. However, the diagonal terms, in which both

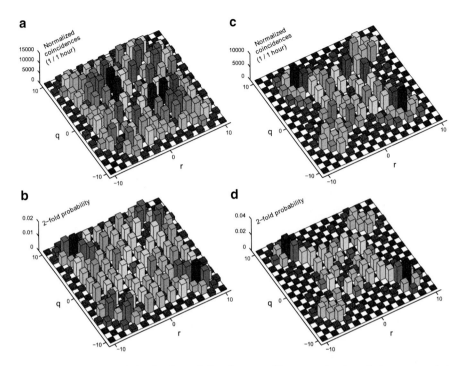

**Fig. 3.96** Measured and simulated correlations in waveguide arrays when two photons are coupled to waveguides $-1$ and 1, (**a** and **b**) for input photons separated with temporal delay longer than their coherence length and (**c** and **d**) for the photons arriving simultaneously in the array (From Peruzzo et al. (2010))

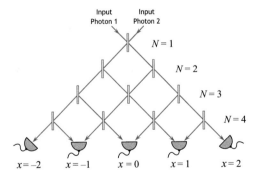

**Fig. 3.97** Outline of a simple optical setup for performing a two-particle coined quantum walk on a line (Adapted from Rohde et al. (2011))

photons appear at the same output (the meeting problem), inherently require number resolving detection.

A physical realisation of Rohde et al.'s (2011) proposed setup in Fig. 3.97 was demonstrated by Sansoni et al. (2012) who utilised an integrated waveguide

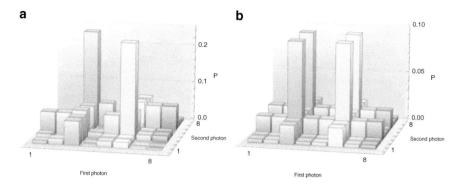

**Fig. 3.98** Simulated joint probability distributions for a pair of indistinguishable (**a**) and distinguishable (**b**) quantum walkers after a four step quantum walk. Grid element $\{m, n\}$ corresponds to the term with a photon at spatial mode $x = m$ and a photon at mode $x = n$ (From Rohde et al. (2011))

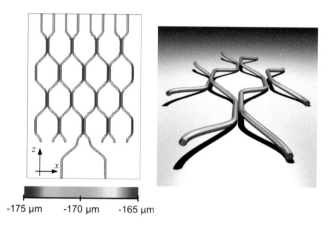

**Fig. 3.99** Directional couplers fabricated for implementing a four-step quantum walk (From Sansoni et al. (2012))

architecture to concentrate a large number of optical elements on a small chip, achieving intrinsic phase stability due to the monolithic structure of the device. In this waveguide implementation, beam splitters are replaced by *directional couplers* – structures in which two waveguides are brought close together for a certain interaction length as depicted in Fig. 3.99, acting as a Mach-Zehnder interferometer.

To realize the integrated optical circuits Sansoni et al. (2012) adopted a femtosecond laser writing technology in which nonlinear absorption of focused femtosecond pulses is exploited to induce permanent and localized refractive index increase in transparent materials. Despite the low birefringence of the substrate, the guided modes for the two linear polarizations of the field are still slightly different,

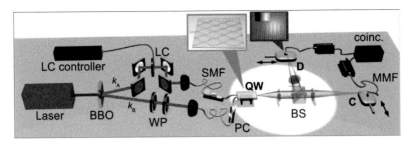

**Fig. 3.100** Experimental setup of an integrated photonics device, including the source, integrated quantum walk circuit, and measuring apparatus (From Sansoni et al. (2012))

resulting in a small polarization dependence in the properties of the fabricated directional couplers. By carefully adjusting the orientation of the waveguides plane with respect to the horizontal however, the authors managed to fabricate a device with polarization insensitive couplers. Furthermore setting the coupling distance to 11 μm and the interaction length to 2.1 mm ensured that the couplers produced a balanced splitting ratio, similar to a beam splitter.

A schematic of the complete experimental setup is shown in Fig. 3.100 including a polarization entangled two-photon source, the 32 mm long integrated optical chip for performing the quantum walk, followed by the measurement unit. The two-photon quantum walk is initialized by generating entangled states of the form

$$|\Psi^\phi\rangle = \frac{1}{\sqrt{2}} \left( |H\rangle_A |V\rangle_B + e^{i\phi} |V\rangle_A |H\rangle_B \right), \qquad (3.167)$$

where the desired phase angle $\phi$ is fixed by the appropriate rotations of half- and quarter-wave plates. Photon pairs are then injected into the integrated optical chip and the emerging distribution reconstructed by measuring the coincidence counts for each combination of output ports corresponding to positions $x_1$ and $x_2$ of the quantum walk.

Figure 3.101 presents the theoretical probability distributions and their experimental counterparts for three different initial states: the triplet state $|\Psi^+\rangle$ with $\phi = 0$, the singlet state $|\Psi^-\rangle$ with $\phi = \pi$, and another generic entangled state with $\phi = \pi/2$. These initial states are particularly significant as they respectively simulate the quantum walk of a pair of bosons, fermions and anions, each exhibiting unique bunching and anti-bunching behaviour.

Most recently Schreiber et al. (2012) demonstrated an intriguing experimental implementation of a two-particle quantum walk on a line using the two-dimensional walk of a single particle. To this end, the authors extended an original scheme introduced in Schreiber et al. (2010) by incorporating an additional degree of freedom in the form of a secondary pathway from the output of PBS$_2$ in Fig. 3.102 back to PBS$_1$ input.

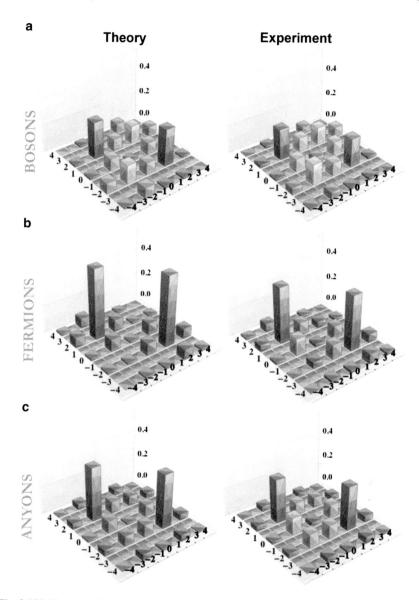

**Fig. 3.101** Two-particle quantum walks: ideal (*left*) and measured (*right*) distributions of (**a**) bosonic ($\phi = 0$), (**b**) fermionic ($\phi = \pi$) and (**c**) anyonic ($\phi = \pi/2$) quantum walks

The coin space of the two-dimensional walk is spanned by four basis states $|c_x, c_y\rangle$ defined by $\{|a\rangle, |b\rangle\} \otimes \{|H\rangle, |V\rangle\}$ where $c_x$ is encoded by the two spatial modes $a$ and $b$ corresponding to the two input ports of $\text{PBS}_1$ and $c_y$ is encoded by $H$ and $V$ linear polarizations. The electro-optic modulator and each of the half-wave plates perform the action of coin operators $\hat{C}_{\text{EOM}}, \hat{C}_{\text{H}_1}, \hat{C}_{\text{H}_2}, \hat{C}_{\text{H}_3}$ and $\hat{C}_{\text{H}_4}$ which can be readily expressed as a $4 \times 4$ matrix. For example

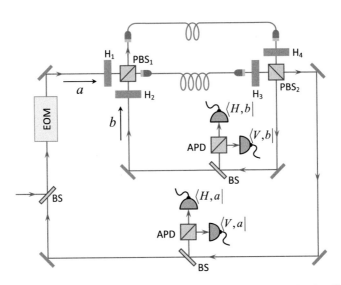

**Fig. 3.102** Schematic diagram of the setup used to perform the two-dimensional walk (Adapted from Schreiber et al. (2012))

$$
\hat{C}_{H_1} = \begin{pmatrix}
\cos(2\theta_1) & \sin(2\theta_1) & 0 & 0 \\
\sin(2\theta_1) & -\cos(2\theta_1) & 0 & 0 \\
0 & 0 & 1 & 0 \\
0 & 0 & 0 & 1
\end{pmatrix}, \tag{3.168}
$$

where $\theta_i$ represents the tilt angle of the half-wave plate relative to the horizontal. Position states of the walk are defined following the same procedure described in Schreiber et al. (2010), i.e. as temporal dispositions of the pulse components travelling around the loop through different paths. More specifically the length difference between the two optic fibers connecting $PBS_1$ to $PBS_2$ introduces a temporal displacement $\delta t_x = 46.42\,\text{ns}$ between the $a$ and $b$ components of the pulse which can be interpreted as conditional steps $|a, x\rangle \longrightarrow |a, x - 1\rangle$ and $|b, x\rangle \longrightarrow |b, x + 1\rangle$ with step size $\delta t_x/2$. Likewise the two paths connecting $PBS_2$ back to $PBS_1$ introduces another, but this time much shorter, temporal displacement $\delta t_y = 3.11\,\text{ns}$ between the $a$ and $b$ components which can be interpreted as conditional *sub*-steps $|a, y\rangle \longrightarrow |a, y - 1\rangle$ and $|b, y\rangle \longrightarrow |b, y + 1\rangle$ with step size $\delta t_y/2$, which are identically applied at every step $x$. This arrangement naturally gives rise to a two-dimensional $(x, y)$ position space shown in Figs. 3.103 and 3.104 where the joint probability distributions are constructed using a pair of time resolving single photon detectors. Note that the relationship between $\delta t_x$ and $\delta t_y$ places an upper bound on the number of possible steps. With $\delta t_x \gtrsim 13 \cdot \delta t_y$ for example, as is the case here, only 12 steps of the walk are possible before the temporal overlap of sub-steps $y$ begins to compromise the walk's integrity.

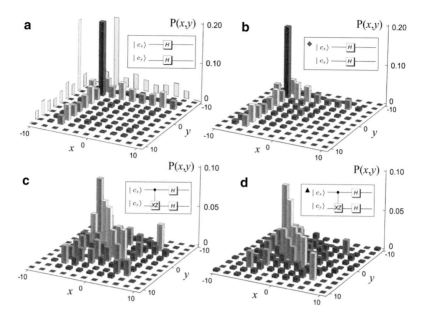

**Fig. 3.103** Theoretical (**a, c**) and measured (**b, d**) probability distribution $P(x, y)$ (traced over the coin space) after ten steps of a two-dimensional quantum walk with initial state $|0, 0, a, H\rangle$. In (**a**) and (**b**), as only separable coin operations were performed (*inset*), the distribution is separable, given by a product of two one dimensional distributions (*gray*). In (**c**) and (**d**) however the coin operations involve a controlled-Not $X$ and controlled-phase operation $Z$, resulting in an unfactorizable distribution. (From Schreiber et al. (2012))

It is now possible to define two coin operators $\hat{C}_x = \hat{C}_{\text{EOM}}\hat{C}_{\text{H}_1}\hat{C}_{\text{H}_2}$ and $\hat{C}_y = \hat{C}_{\text{H}_3}\hat{C}_{\text{H}_4}$ which act on the quantum coin state $|c_x, c_y\rangle$. The key insight gleaned by Schreiber et al. (2012) was that quantum walks driven only by separable coins $\hat{C}_x$ and $\hat{C}_y$ acting in each direction can be factorized into two independent distributions of one dimensional quantum walks, stating no conceptual advantage of a two-dimensional quantum walk. An example of this is shown in Figs. 3.103a and 3.103b for the case where all four half-wave plates are set to perform a Hadamard operation ($\theta_{1,2,3,4} = 22.5°$). Crucially though if the coin operations are modified such that the transformation of one coin state is conditioned on the actual state of the other, then due to the induced quantum correlations one obtains a non-trivial evolution resulting in an inseparable final state. An example of this is shown in Figs. 3.103c and 3.103d for the case where the coin operators have been adapted to perform an additional controlled-Not and controlled-phase operation ($\theta_1 = -22.5°$ and $\theta_{2,3,4} = 22.5°$). The authors noted that each combined coin operation in a two-particle walk, including controlled operations, has an equivalent coin operation in a two-dimensional quantum walk. This allows one to interpret the above results as a quantum walk of two virtual particles on a line with controlled two-particle operations, a system typically creating two-particle entanglement. The

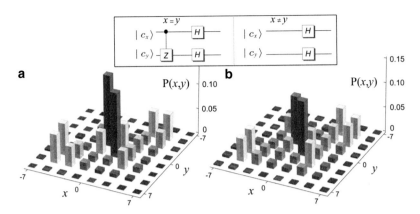

**Fig. 3.104** Theoretical (**a**) and measured (**b**) coincidence distribution $P(x, y)$ (traced over the coin space) after seven steps of a simulated two-particle quantum walk with initial state $|0, 0, a, H\rangle$ and non-linear interactions (*inset*) (From Schreiber et al. (2012))

inseparability of the final probability distribution is then a direct signature of the simulated entanglement.

Schreiber et al. (2012) further noted that interactions introduced using controlled, but static coin operations can be interpreted as long distance since the interaction strength is independent of the spatial distance of the particles. This however is a rather unique effect and highly non-trivial to demonstrate in actual two-particle quantum systems. Short-range interactions on the other hand occur only when both particles occupy the same position and can be interpreted as two-particle scattering or non-linear interactions. Schreiber et al. (2010) utilized the EOM to simulate such a position dependant short range interaction as shown in Fig. 3.104.

The experiment of Schreiber et al. (2012) is a remarkable demonstration of how entanglement of virtual particles may be simulated using essentially classical optics. At the same time such implementation schemes further highlight the exponential trade off associated with the number of required classical resources, rendering an inherently multi-particle implementation highly attractive.

# Abstract

Given the extensive applications of random walks to classical algorithms in virtually every science related discipline, we may be at the threshold of yet another problem solving paradigm with the advent of quantum walks. Over the past decade, quantum walks have been extensively explored for their non-intuitive dynamics, which may hold the key to radically new quantum algorithms. This growing interest in the theoretical applications of quantum walks has been paralleled by a flurry of research into a more practical problem: how does one physically implement a quantum walk in the laboratory? This book provides a comprehensive survey of numerous proposals, as well as actual experiments, for a physical realisation of quantum walks, underpinned by a wide range of quantum, classical and hybrid technologies.

J. Wang and K. Manouchehri, *Physical Implementation of Quantum Walks*,
Quantum Science and Technology, DOI 10.1007/978-3-642-36014-5,
© Springer-Verlag Berlin Heidelberg 2014

# Appendix A
# Electromagnetic Radiation

## A.1 Classical Radiation Field

*Main References (Townsend 2000b; Shore 1990a; Griffiths 1999)*

Classically, the electromagnetic field represents the general solution to Maxwell's equations which can be expressed in terms of a scalar potential $\varphi$ and a vector potential $\mathbf{A}$, whereby for some charge density $\rho$ and volume current density vector $\mathbf{j}$

$$
\begin{cases}
\mathbf{E} = -\nabla\varphi - \dfrac{\partial \mathbf{A}}{\partial t}, \\[6pt]
\mathbf{B} = \nabla \times \mathbf{A}, \\[6pt]
-\nabla^2\varphi + \dfrac{\partial}{\partial t}(\nabla \cdot \mathbf{A}) = -\dfrac{1}{\epsilon_0}\rho, \\[6pt]
\left(\nabla^2 \mathbf{A} - \mu_0\epsilon_0 \dfrac{\partial^2 \mathbf{A}}{\partial t^2}\right) - \nabla\left(\nabla \cdot \mathbf{A} + \mu_0\epsilon_0 \dfrac{\partial\varphi}{\partial t}\right) = -\mu_0\mathbf{j},
\end{cases}
\tag{A.1}
$$

with $\epsilon_0$ and $\mu_0$ representing the electric and magnetic vacuum permittivity respectively. Using the Coulomb gauge condition

$$
\nabla \cdot \mathbf{A} = 0,
\tag{A.2}
$$

and in the absence of charges and currents (i.e. $\rho = \mathbf{j} = 0$), the scalar potential $\varphi = 0$ and the vector potential $\mathbf{A}$ satisfies the wave equation

$$
\frac{1}{c^2}\frac{\partial^2 \mathbf{A}}{\partial t^2} - \nabla^2 \mathbf{A} = 0,
\tag{A.3}
$$

for which the generalized traveling plane wave solution is given by

$$
\mathbf{A}(\mathbf{r}, t) = \sum_{\lambda=1,2} \int d\mathbf{k}\, d\nu \left( A_{\mathbf{k},\nu,\lambda}\, \mathbf{u}(\mathbf{k}, \nu, \lambda)\, e^{i(\mathbf{k}\cdot\mathbf{r} - \omega_{\mathbf{k},\nu} t)} + \text{c.c.} \right),
\tag{A.4}
$$

J. Wang and K. Manouchehri, *Physical Implementation of Quantum Walks*,
Quantum Science and Technology, DOI 10.1007/978-3-642-36014-5,
© Springer-Verlag Berlin Heidelberg 2014

where c.c. denotes "complex conjugate", $\mathbf{k}$ labels the spatial modes of radiation propagating along vector $\mathbf{k}$, $\nu$ labels the frequency modes, $\lambda$ labels the two linearly independent (real valued) polarization unit vectors $\mathbf{u}$ such that $\mathbf{k} \cdot \mathbf{u}(\mathbf{k}, \nu, \lambda) = 0$ (satisfying Eq. A.2) and $A_{\mathbf{k},\nu,\lambda}$ is a complex scalar. In addition, for a valid solution, the wave vector and the angular frequency are not independent and must adhere to the dispersion relation

$$|\mathbf{k}| = \frac{\omega_{\mathbf{k},\nu}}{c}. \tag{A.5}$$

The electric component of the electromagnetic field is then given by

$$\begin{aligned}
\mathbf{E}(\mathbf{r}, t) &= -\frac{\partial \mathbf{A}}{\partial t} \\
&= \sum_{\lambda=1,2} \frac{1}{2} \int d\mathbf{k}\, d\nu \left( E_{\mathbf{k},\nu,\lambda}\, \mathbf{u}(\mathbf{k}, \nu, \lambda)\, e^{i(\mathbf{k}\cdot\mathbf{r}-\omega_{\mathbf{k},\nu}t)} + \text{c.c.} \right) \\
&= \sum_{\lambda=1,2} \int d\mathbf{k}\, d\nu\, \mathbf{E}_{\mathbf{k},\nu,\lambda}\, \cos(\mathbf{k}\cdot\mathbf{r} - \omega_{\mathbf{k},\nu}t - \phi_{\mathbf{k},\nu,\lambda}), \tag{A.6}
\end{aligned}$$

where

$$E_{\mathbf{k},\nu,\lambda} = 2i\,\omega_{\mathbf{k},\nu} A_{\mathbf{k},\nu,\lambda} = |E_{\mathbf{k},\nu,\lambda}|\, e^{i\phi_{\mathbf{k},\nu,\lambda}}, \tag{A.7}$$

and the real valued vector

$$\mathbf{E}_{\mathbf{k},\nu,\lambda} = |E_{\mathbf{k},\nu,\lambda}|\, \mathbf{u}(\mathbf{k}, \nu, \lambda). \tag{A.8}$$

It is instructive to note that when interacting with atoms and molecules, the magnetic component $\mathbf{B}(\mathbf{r}, t)$ of the radiation field is often neglected, due to much weaker interaction energy compared with the electric field (see Appendix B.1.3).

A commonly used specialization of the generalized electric field in Eq. A.4 is the single mode radiation traveling in direction $z$, for which

$$\mathbf{E}(z, t) = \frac{1}{2}\mathbf{E}^{(+)}(z, t) + \frac{1}{2}\mathbf{E}^{(-)}(z, t) \tag{A.9}$$

$$= \frac{1}{2}\mathbf{J}e^{i(kz-\omega t)} + \frac{1}{2}\mathbf{J}^* e^{-i(kz-\omega t)} \tag{A.10}$$

$$\equiv \frac{1}{2}\mathbf{e}E_0 e^{i(kz-\omega t)} + \frac{1}{2}\mathbf{e}^* E_0 e^{-i(kz-\omega t)} \tag{A.11}$$

$$\equiv \frac{1}{2}\mathbf{E}^{(+)}(z)e^{-i\omega t} + \frac{1}{2}\mathbf{E}^{(-)}(z)e^{i\omega t} \tag{A.12}$$

$$\equiv \frac{1}{2}\mathbf{E}^{(+)}(\omega)e^{ikz} + \frac{1}{2}\mathbf{E}^{(-)}(\omega)e^{-ikz}, \tag{A.13}$$

where the complex vector

$$\mathbf{J} = \begin{pmatrix} E_x e^{i\phi_x} \\ E_y e^{i\phi_y} \end{pmatrix} \tag{A.14}$$

is known as the *Jones vector*, often normalized to a unit vector $\mathbf{e}$, $E_x$ and $E_y$ are real-valued amplitudes, $E_0 = \sqrt{E_x^2 + E_y^2}$ and the last two lines can be used to explicate either the temporal or the spatial parameters of the field. This electric field is said to be linearly polarized when $\phi_x = \phi_y = \phi$ which yields the familiar case

$$\mathbf{E}(z,t) = \mathbf{E}_0 \cos(\omega t - kz - \phi), \tag{A.15}$$

where $\mathbf{E}_0 = (E_x, E_y)$, corresponding to

$$\mathbf{e} = \frac{1}{\sqrt{2}} \begin{pmatrix} 1 \\ 1 \end{pmatrix}, \tag{A.16}$$

for $E_x = E_y$. To obtain a circularly polarized radiation on the other hand, we require $|\phi_x - \phi_y| = \pm\pi/2$ and $E_x = E_y$, corresponding to right- and left-handed orthonormal vectors

$$\mathbf{e}_\pm = \frac{1}{\sqrt{2}} \begin{pmatrix} 1 \\ \pm i \end{pmatrix}, \tag{A.17}$$

assuming $\phi_x = 0$.

## A.2   Quantized Radiation Field

*Main Reference (Townsend 2000b)*

A quantum description of light is obtained by quantizing the radiation field. A standard way to do this is to consider the radiation in a cubic cavity with an arbitrary volume $V = L^3$ and the periodic boundary condition

$$e^{ik_j j} = e^{ik_j(j+L)}, \tag{A.18}$$

for $j = x, y, z$. This leads to the quantization of the wave vector according to

$$k_j = \frac{2\pi n_j}{L}, \tag{A.19}$$

where $n_j = 0, \pm 1, \pm 2, \dots$ which in turn requires frequency modes to also be discrete due to the dispersion relation. Note that the introduction of volume $V$ is purely for mathematical convenience and does not appear in any of the observables. Now considering the classical electromagnetic field energy of a system confined to volume $V$

$$
\begin{aligned}
H_{\text{EM}} &= \frac{1}{2} \int_V d^3 r \, \left( \epsilon_0 \mathbf{E}_0^2 + \frac{1}{\mu_0} \mathbf{B}_0^2 \right) \\
&= \frac{1}{2} \int_V d^3 r \, \left\{ \epsilon_0 \left( -\frac{\partial \mathbf{A}}{\partial t} \right)^2 + \frac{1}{\mu_0} (\nabla \times \mathbf{A})^2 \right\} ,
\end{aligned}
\tag{A.20}
$$

vector $\mathbf{A}$ can be represented using a discretized form of Eq. A.4 given by

$$
\mathbf{A}(\mathbf{r}, t) = \sum_{\mathcal{M}} \left( A_{\mathcal{M}} \, \mathbf{u}(\mathcal{M}) \, \frac{e^{i(\mathbf{k} \cdot \mathbf{r} - \omega_{\mathbf{k},\nu} t)}}{\sqrt{V}} + \text{c.c.} \right),
\tag{A.21}
$$

where the shorthand notation $\mathcal{M} = \{\mathbf{k}, \nu, \lambda\}$ identifies a given mode with a unique set of spatial, frequency and polarization parameters. Taking advantage of the orthonormality relation

$$
\int_V d^3 r \, \frac{e^{i\mathbf{k} \cdot \mathbf{r}}}{\sqrt{V}} \frac{e^{-i\mathbf{k}' \cdot \mathbf{r}}}{\sqrt{V}} = \delta_{\mathbf{k}\mathbf{k}'},
\tag{A.22}
$$

it is possible to show that

$$
\begin{aligned}
H_{\text{EM}} &= 2\epsilon_0 \sum_{\mathcal{M}} \omega_{\mathcal{M}}^2 \, A_{\mathcal{M}}^* A_{\mathcal{M}} \\
&= \sum_{\mathcal{M}} \left( \frac{p_{\mathcal{M}}^2}{2} + \frac{1}{2} \omega_{\mathcal{M}}^2 q_{\mathcal{M}}^2 \right),
\end{aligned}
\tag{A.23}
$$

where

$$
p_{\mathcal{M}} = -i\omega_{\mathcal{M}} \sqrt{\epsilon_0} (A_{\mathcal{M}} - A_{\mathcal{M}}^*),
\tag{A.24}
$$

$$
q_{\mathcal{M}} = \sqrt{\epsilon_0} (A_{\mathcal{M}} + A_{\mathcal{M}}^*).
\tag{A.25}
$$

The above equation closely resembles an ensemble of harmonic oscillators for which the local energy at any one infinitesimal point in space can be determined quantum mechanically via the usual operator

$$
\hat{H}_{\text{HO}} = \sum_{s=x,y,z} \frac{\hat{p}_s^2}{2m} + \frac{1}{2} m \omega^2 \hat{s}^2,
\tag{A.26}
$$

where for $m = 1$ we have

$$\hat{p}_s = -i\sqrt{\frac{\omega\hbar}{2}}(\hat{a}_s - \hat{a}_s^\dagger), \tag{A.27}$$

$$\hat{s} = \sqrt{\frac{\hbar}{2\omega}}(\hat{a}_s + \hat{a}_s^\dagger), \tag{A.28}$$

and $\hat{a}_s$ and $\hat{a}_s^\dagger$ are the annihilation and creation operators for mode $s$. This enables us to write equivalent quantum mechanical operators for the classical field components. Making the substitutions

$$A_\mathcal{M} \longrightarrow \sqrt{\frac{\hbar}{2\epsilon_0\omega_\mathcal{M}}}\hat{a}_\mathcal{M}, \tag{A.29}$$

$$A_\mathcal{M}^* \longrightarrow \sqrt{\frac{\hbar}{2\epsilon_0\omega_\mathcal{M}}}\hat{a}_\mathcal{M}^\dagger, \tag{A.30}$$

the vector potential in Eq. A.4 can then be transformed to an operator

$$\hat{\mathbf{A}}(\mathbf{r}, t) = \sum_\mathcal{M} \sqrt{\frac{\hbar}{2\epsilon_0\omega_\mathcal{M}V}}\, \mathbf{u}(\mathcal{M})\left(\hat{a}_\mathcal{M}\, e^{i(\mathbf{k}\cdot\mathbf{r} - \omega_\mathcal{M}t)} + \text{c.c.}\right). \tag{A.31}$$

Using this expression to evaluate the Hamiltonian in Eq. A.20, which is now also an operator, we obtain

$$\hat{H}_{\text{EM}} = \frac{1}{2}\sum_\mathcal{M} \hbar\omega_\mathcal{M}\left(\hat{a}_\mathcal{M}\hat{a}_\mathcal{M}^\dagger + \hat{a}_\mathcal{M}^\dagger\hat{a}_\mathcal{M}\right)$$

$$= \sum_\mathcal{M} \hbar\omega_\mathcal{M}\left(\hat{a}_\mathcal{M}\hat{a}_\mathcal{M}^\dagger + \frac{1}{2}\right), \tag{A.32}$$

where the last step takes advantage of the communication relations

$$[\hat{a}_\mathcal{M}, \hat{a}_{\mathcal{M}'}^\dagger] = \delta_{\mathcal{M},\mathcal{M}'}. \tag{A.33}$$

Equation A.32 represents an ensemble of the familiar harmonic oscillator Hamiltonians. The total field energy in the box is determined by the observable $\langle\psi|\hat{H}_{\text{EM}}|\psi\rangle$, where $|\psi\rangle$ represents the quantum state of the field over the entire cavity volume $V$ that can be made arbitrarily large to encapsulate the entire quantum system under investigation. We can also derive the electric field operator

$$\hat{\mathbf{E}}(\mathbf{r}, t) = -\frac{\partial \hat{\mathbf{A}}}{\partial t}$$

$$= \sum_{\mathbf{k}, v, \lambda} i \; \eta_{\mathbf{k}, v, \lambda} \; \mathbf{u} \; \hat{a}_{\mathbf{k}, v, \lambda}(\mathbf{k}, \lambda) \; e^{i(\mathbf{k}\cdot\mathbf{r} - \omega_{\mathbf{k}, v} t)} + \text{c.c.,} \qquad (A.34)$$

where

$$\eta_{\mathbf{k}, v, \lambda} = \sqrt{\frac{\hbar \omega_{\mathbf{k}, v}}{2\epsilon_0 V}} \qquad (A.35)$$

is the quantum mechanical unit of electric field. Subsequently the operator for the single mode field in Eq. A.9 is given by

$$\hat{\mathbf{E}}(z, t) = \frac{1}{2} \left( \hat{\mathbf{E}}^{(+)}(z) e^{-i\omega t} + \hat{\mathbf{E}}^{(-)}(z) e^{i\omega t} \right), \qquad (A.36)$$

where $\hat{\mathbf{E}}^{(+)}(z) = i \eta \mathbf{u} \, \hat{a} \, e^{ikz}$.

As evident from the above discussion, particularly Eq. A.32, each unique radiation mode in the quantum field is modeled by a simple harmonic oscillator. To complete the quantization, we now represent each oscillator by a Fock state $|n\rangle$, with $n = 0, 1, 2, \ldots$ being the number of photons in that state. The overall quantized radiation field for a specific value of $n$ is hence given by

$$|n_{\mathcal{M}_1}\rangle |n_{\mathcal{M}_2}\rangle |n_{\mathcal{M}_3}\rangle \ldots = \prod_{\mathcal{M}} |n_{\mathcal{M}}\rangle \equiv |\{n_{\mathcal{M}}\}\rangle. \qquad (A.37)$$

Any arbitrary pure state can now be constructed as a superposition of the basis states, i.e.

$$|\text{pure state}\rangle = \sum_i c_i |\{n_{i,\mathcal{M}}\}\rangle$$

$$= \sum_i c_i |n_{i,\mathcal{M}_1}\rangle |n_{i,\mathcal{M}_2}\rangle |n_{i,\mathcal{M}_3}\rangle \ldots, \qquad (A.38)$$

where $c_i$ is a complex coefficient and $n_i = 0, 1, 2, \ldots \infty$.

## A.3   Coherent Versus Fock States

*Main Reference (Glauber 1963)*

A single mode radiation field is said to be in a coherent or Glauber state (Glauber 1963) when it is represented by

$$|\alpha\rangle = e^{-|\alpha|^2/2} \sum_n \frac{\alpha^n}{\sqrt{n!}} |n\rangle, \tag{A.39}$$

where $\alpha$ is a dimensionless complex number, and $|\alpha|^2 = \langle n \rangle \equiv \langle \hat{a}^*_{\mathcal{M}} \hat{a}_{\mathcal{M}} \rangle$ represents the average photon number in the mode. In contrast to Fock states however, coherent states are non-orthogonal. Evaluating the wave-function in the Fock space we find

$$P(n) = |\psi_\alpha(n)|^2 = |\langle n|\alpha\rangle|^2 = e^{-\langle n \rangle} \frac{\langle n \rangle^n}{n!}, \tag{A.40}$$

meaning that the photon number in the cavity follows a Poissonian distribution (Fig. A.1c). In the real space on the other hand

$$\psi_\alpha(x) = \langle x|\alpha\rangle = e^{-|\alpha|^2/2} \sum_n \frac{\alpha^n}{\sqrt{n!}} \psi_n(x), \tag{A.41}$$

where position space energy eigenfunctions

$$\psi_n(x) = \langle x|n\rangle = \frac{1}{\sqrt{n!}} \left(\sqrt{\frac{\kappa}{2}}\right)^n \left(x - \frac{1}{\kappa}\frac{d}{dx}\right)^n \left(\frac{\kappa}{\pi}\right)^{1/4} e^{-\kappa x^2/2}, \tag{A.42}$$

and $\kappa = m\omega/\hbar$. By adopting implicit units $\kappa \longrightarrow 1$ these eigenfunctions can be more conveniently represented using Hermite polynomials

$$\psi_n(x) = \frac{1}{\sqrt{\sqrt{\pi} 2^n n!}} H_n(x) e^{-x^2/2}, \tag{A.43}$$

depicted in Fig. A.1a.

Another form in which coherent states may be formulated is

$$|\alpha\rangle = e^{\alpha \hat{a}^\dagger - \alpha^* \hat{a}} |0\rangle = \hat{D}(\alpha)|0\rangle,$$

where $\hat{D}$ is the displacement operator (Glauber 1963). Expressing the displacement operator in real space for a real valued $\ell$ we obtain

$$\hat{D}(\ell) = e^{\ell(\hat{a}^\dagger - \hat{a})}$$

$$= e^{-\ell\sqrt{\frac{2\hbar}{m\omega}}\frac{d}{dx}}$$

$$= e^{-\delta\frac{d}{dx}}, \tag{A.44}$$

**Fig. A.1** (**a**) A plot of
Eq. A.43, the real space
wave-function $\psi_n(x)$ for
Fock states $|n\rangle$ with photon
number $n = 0$ (*solid*), $n = 1$
(*dotted*), and $n = 2$ (*dashed*).
Superposition of Fock states
according to Eq. A.39
produces a coherent state. (**b**)
The real space wave-function
$\psi_\alpha(x)$ for the coherent states
$|\alpha\rangle$ with $\alpha = 2$ (*solid*) and
$\alpha = 4$ (*dashed*). Position $x$ is
in units of $\sqrt{2\hbar/m\omega}$. (**c**)
Photon number probability
distribution $P(n) = |\psi_\alpha(n)|^2$
in the Fock space for the
coherent states with $\alpha = 2$
(*solid*) and $\alpha = 4$ (*dashed*).
The distribution is a
Poissonian described by
Eq. A.40

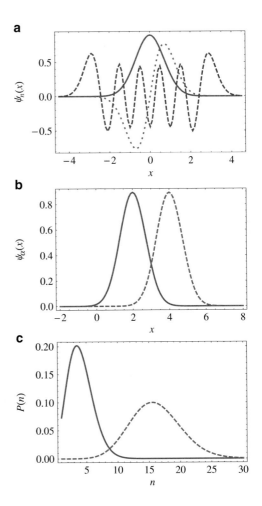

where $\delta = \ell\sqrt{\frac{2\hbar}{m\omega}}$ is a dimensionless length. Hence, for small $\delta$ the action of the of
the displacement operator on an arbitrary function

$$\hat{D}f(x) = e^{-\delta\frac{d}{dx}}f(x)$$

$$= (1 - \delta\frac{d}{dx} + \frac{1}{2}\delta^2\frac{d^2}{dx^2} + \ldots)f(x)$$

$$= f(x - \delta) \tag{A.45}$$

formally displaces the function by a length $\delta$ in real space. As illustrated in
Fig. A.1b, the wave-function $\psi_\alpha(x)$ with real $\alpha$ describes a wave-packet with a
finite width which corresponds to the photon distribution inside the physical space

of the cavity interior. Hence, the wave-function $\psi_{\alpha+\ell}(x)$, where $\ell$ is a real number, describes a spatial displacement, in units of $\sqrt{\frac{2\hbar}{m\omega}}$, of the photon distribution $\psi_\alpha(x)$ by a distance $\ell$ inside the cavity.

## A.4 Reproducing the Classical Result

Of all the states represented by Eq. A.38, one special pure state finds the form

$$|\psi\rangle = \sum_i c_i \sum_j c_j \sum_k c_k \ldots |n_{i,\mathcal{M}_1}\rangle|n_{j,\mathcal{M}_2}\rangle|n_{k,\mathcal{M}_3}\rangle \ldots, \tag{A.46}$$

where $c_i = \alpha^n e^{-|\alpha|^2/2}/\sqrt{n_i!}$. Comparing this with Eq. A.39 we can now see the relationship with a coherent state, from which we obtain

$$|\psi\rangle = \sum_j c_j \sum_k c_k \ldots |\alpha_{\mathcal{M}_1}\rangle|n_{j,\mathcal{M}_2}\rangle|n_{k,\mathcal{M}_3}\rangle \ldots$$

$$= \sum_k c_k \ldots |\alpha_{\mathcal{M}_1}\rangle|\alpha_{\mathcal{M}_2}\rangle|n_{k,\mathcal{M}_3}\rangle \ldots$$

$$\vdots$$

$$= |\alpha_{\mathcal{M}_1}\rangle|\alpha_{\mathcal{M}_2}\rangle|\alpha_{\mathcal{M}_3}\rangle \ldots, \tag{A.47}$$

where $|\alpha_\mathcal{M}|^2 = \langle n_\mathcal{M}\rangle \equiv \langle \hat{a}_\mathcal{M}^* \hat{a}_\mathcal{M}\rangle$ is the mean photon number in mode $\mathcal{M}$. Using the identity $\hat{a}_\mathcal{M}|\alpha_\mathcal{M}\rangle = \alpha_\mathcal{M}|\alpha_\mathcal{M}\rangle$ we can compute the expectation value of the electric field for a coherent state

$$\langle \hat{E}\rangle \equiv \langle\psi|\hat{E}|\psi\rangle, \tag{A.48}$$

which is found to be identical to Eq. A.6 with $E_{\mathbf{k},v,\lambda} = i\,\eta_{\mathbf{k},v,\lambda}\alpha_{\mathbf{k},v,\lambda}$. Furthermore using the identity $\left[\hat{a}_{\mathbf{k},v,\lambda}, \hat{a}_{\mathbf{k}',v',\lambda'}^\dagger\right] = \delta_{\mathbf{k},\mathbf{k}'}\delta_{v,v'}\delta_{\lambda,\lambda'}$ allows us to calculate the uncertainty in the measurement of the electric field

$$\Delta E = \sqrt{\langle\psi|\hat{\mathbf{E}}\cdot\hat{\mathbf{E}}^*|\psi\rangle - \langle\psi|\hat{\mathbf{E}}|\psi\rangle\langle\psi|\hat{\mathbf{E}}^*|\psi\rangle} = \sum_{\mathbf{k},v,\lambda} \eta_{\mathbf{k},v,\lambda}, \tag{A.49}$$

which is a constant. Hence for large photon numbers, coherent states provide the best description of the state of a laser beam. This is demonstrated in Fig. A.2 where the expectation value of the electric field for a single mode coherent state is plotted for different values of $\alpha$.

**Fig. A.2** Expectation value $\langle \hat{E} \rangle \equiv \langle \alpha | \hat{E} | \alpha \rangle$ for a single mode coherent state $|\alpha\rangle$ with (**a**) $\alpha = 1$, (**b**) $\alpha = 2$ and (**c**) $\alpha = 10$

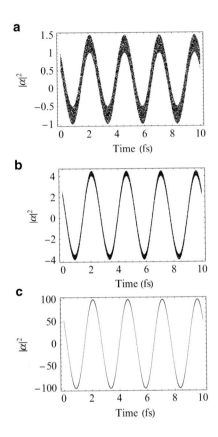

## A.5   Orbital Angular Momentum

*Main Reference (Allen et al. 1992)*

Orbital Angular Momentum (OAM) arises form the Laguerre-Gaussian solution to the wave equation (Eq. A.3) to which the familiar plane wave (Eq. A.4) is another solution. The Laguerre-Gaussian solutions are represented by the basis functions

$$\text{LG}_p^l(r, \phi, z = 0) = \sqrt{\frac{2p!}{\pi(|l| + p)!}} \frac{1}{w} \left( \frac{\sqrt{2}r}{w} \right)^{|l|} L_p^{|l|} \left( \frac{2r^2}{w^2} \right) e^{-r^2/w^2} e^{-il\phi},$$

(A.50)

where $\text{LG}_p^l$ represents a Laguerre-Gaussian mode, $r$ and $\phi$ are the polar coordinates, $L_p^{|l|}$ denotes the associated Laguerre polynomial, $w$ is the radius of the laser beam (at the beam waist located at $z = 0$ along the $z$-axis), the index $l$ is referred to as the winding number and $p$ is the number of nonaxial radial nodes (for the complete form where $z \neq 0$ see Allen et al. (1992) and Vaziri et al. (2002)). Figure A.3 represents the cross-sectional density patterns for some LG modes.

**Fig. A.3** Cross-sectional density profiles for some Laguerre-Gaussian modes denoted by $LG_p^l$: (**a**) $\left|LG_0^0\right|$, (**b**) $\left|LG_0^3\right|$, (**c**) $\left|LG_3^3\right|$ and (**d**) $\left|LG_3^3 + LG_3^{-3}\right|$

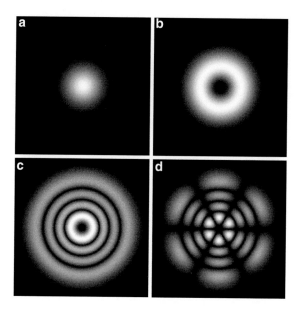

It should be noted that the angular momentum carried by LG modes with $l \neq 0$ is an external angular momentum distinct from the internal angular momentum of the photons associated with their intrinsic helicity. Right and left handed circularly polarized light each carry a single unit of spin angular momentum (SAM) $\pm\hbar$ which render photons a natural candidate for use as qubits. On the other hand LG modes have an azimuthal angular dependency of $\exp(-il\phi)$, where $l$ is the azimuthal mode index and each mode is an eigenmode of the angular momentum operator $L_z$ and carries an orbital angular momentum of $l\hbar$ per photon (Allen et al. 1992). Since there is no upper limit on $l$ photons can also be used as qudit (i.e. multi-valued quantum bits).

# Appendix B
# Atom Optics

Atomic excitations due to the presence of an electromagnetic radiation can often be studied with adequate precision using semi-classical models, most notably the Rabi model for single photon transitions and the Raman model for two-photon transitions, in which the incident light interacting with the atom is described using a classical field. A more accurate and fully quantum mechanical treatment of the subject can however be obtained using the Jaynes-Cummings model and its multi-photon extensions. In this approach quantum operators describe the interaction of quantized electromagnetic radiation with the atomic wave-function.

## B.1 Atom-Field Interaction

*Main Reference (Shore 1990b)*

### B.1.1 Interaction Energy

As a first step in quantifying the atom-field interaction, we consider the energy associated with a charge $q$ placed at point $\mathbf{r}$ relative to the atomic center positioned at $\mathbf{R}$. This is equivalent to the work done by the Lorentz force

$$\mathbf{F} = -q(\mathbf{E} + \mathbf{v} \times \mathbf{B}), \tag{B.1}$$

in moving an initially free charge $q$ first to the stationary coordinate $\mathbf{R}$ and then to location $\mathbf{r}$, as depicted in Fig. B.1. Hence the total energy associated with the charge can be represented as

$$V(\mathbf{r}, t) = V(\mathbf{R}, t) + V_{\text{elec}}(\mathbf{r}, t) + V_{\text{magn}}(\mathbf{r}, t), \tag{B.2}$$

J. Wang and K. Manouchehri, *Physical Implementation of Quantum Walks*,
Quantum Science and Technology, DOI 10.1007/978-3-642-36014-5,
© Springer-Verlag Berlin Heidelberg 2014

**Fig. B.1** The atomic
coordinate system

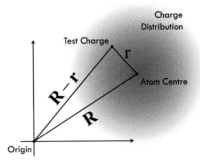

where $V(\mathbf{R}, t)$ is the electrostatic potential energy due to interaction with the atomic nucleus, and $V_{\text{elec}}$ and $V_{\text{magn}}$ correspond to the contributions of the electric and magnetic terms in the Lorentz force. The energy due to the electric field can be written as

$$
\begin{aligned}
V_{\text{elec}}(\mathbf{r}, t) &= -q \int_0^{\mathbf{r}} d\mathbf{s} \cdot \mathbf{E}(\mathbf{R} + \mathbf{s}, t) \\
&= -q \int_0^1 d\lambda\, \mathbf{r} \cdot \mathbf{E}(\mathbf{R} + \lambda \mathbf{r}, t) \\
&= -q \sum_{n=1}^{\infty} \frac{1}{n!} \left( \mathbf{r} \cdot \frac{\partial}{\partial \mathbf{R}} \right)^{n-1} \mathbf{r} \cdot \mathbf{E}(\mathbf{R}, t),
\end{aligned}
\tag{B.3}
$$

where in deriving the second step we have replaced the displacement $\mathbf{s}$ with $\lambda \mathbf{r}$ for $\lambda \in [0, 1]$, and then carried out the integration over $\lambda$ using the vector form of the 1D Taylor expansion

$$
\begin{aligned}
f_i(\mathbf{R} + \mathbf{s}) &= -\sum_{n=0}^{\infty} \frac{1}{n!} \left( \sum_{j=1}^{3} s_j \frac{\partial}{\partial R_j} \right)^n f_i(\mathbf{R}) \\
&= -\sum_{n=0}^{\infty} \frac{1}{n!} \left( \mathbf{s} \cdot \frac{\partial}{\partial \mathbf{R}} \right)^n f_i(\mathbf{R})
\end{aligned}
\tag{B.4}
$$

with $i$ and $j$ representing the three Cartesian components of the vector. Likewise, the energy associated with a classical charge inside a magnetic field is given by

$$
\begin{aligned}
V_\ell(\mathbf{r}, t) &= -q \int_0^{\mathbf{r}} d\mathbf{s} \cdot \dot{\mathbf{r}} \times B(\mathbf{R} + \mathbf{s}, t) \\
&= -q \sum_{n=1}^{\infty} \frac{1}{(n+1)!} \left( \mathbf{r} \cdot \frac{\partial}{\partial \mathbf{R}} \right)^{n-1} (\mathbf{r} \times \dot{\mathbf{r}}) \cdot \mathbf{B}(\mathbf{R}, t) \\
&= -\frac{q\hbar}{m} \sum_{n=1}^{\infty} \frac{1}{(n+1)!} \left( \mathbf{r} \cdot \frac{\partial}{\partial \mathbf{R}} \right)^{n-1} \boldsymbol{\ell} \cdot \mathbf{B}(\mathbf{R}, t),
\end{aligned}
\tag{B.5}
$$

where we have additionally used the identity $\mathbf{a} \cdot (\mathbf{b} \times \mathbf{c}) \equiv (\mathbf{a} \times \mathbf{b}) \cdot \mathbf{c}$ in arriving at the second line. This is then followed by introducing the definition $\hbar \boldsymbol{\ell} = r \times p \simeq \mathbf{r} \times m\dot{\mathbf{r}}$, where $\boldsymbol{\ell}$ is a dimensionless angular momentum vector, $m$ is the particle mass and $\mathbf{p} = m\mathbf{v}$ neglects the diamagnetic and relativistic effects. Vector $\boldsymbol{\ell}$ arises naturally from the quantum mechanical treatment of a particle's probability wave-function but its magnitude is quantized. Considering the electrons in an atom for example, $\hbar \boldsymbol{\ell} \equiv \mathbf{L}$ gives the orbital angular momentum vector and $|\boldsymbol{\ell}|$ can only assume integer values $0, 1, 2 \ldots$ which are loosely associated with orbitals $s, p, d \ldots$ of the atom.

Another property assumed by particles in the quantum regime is an intrinsic spin $\mathbf{s}$ expressed in units of $\hbar$, where $|\mathbf{s}|$ is either an integer or half an odd integer. Vector $\mathbf{s}$ is analogous in behavior to an angular momentum vector and particles with non-zero spin possess an intrinsic magnetic moment expressed as

$$\boldsymbol{\mu}_s = -\frac{q\hbar}{2m} g_s \mathbf{s}. \tag{B.6}$$

For electrons $|\mathbf{s}| = 1/2$ and the dimensionless g-factor $g_s \simeq 2$. The orientation energy of the intrinsic magnetic moment $\boldsymbol{\mu}_s$ in a magnetic field is given by

$$V_s(\mathbf{r}, t) = -\int_0^\infty d\lambda \boldsymbol{\mu}_s \cdot B(\mathbf{R} + \lambda \mathbf{r}, t)$$

$$= -\frac{q\hbar}{2m} g_s \sum_{n=1}^\infty \frac{1}{n!} \left( \mathbf{r} \cdot \frac{\partial}{\partial \mathbf{R}} \right)^{n-1} \mathbf{s} \cdot \mathbf{B}(\mathbf{R}, t). \tag{B.7}$$

The total contribution of the magnetic field can then be expressed as

$$V_{\text{magn}}(\mathbf{r}, t) = V_\ell(\mathbf{r}, t) + V_s(\mathbf{r}, t) \tag{B.8}$$

$$V_{\text{magn}}(\mathbf{r}, t) = -\frac{q\hbar}{2m} \sum_{n=1}^\infty \frac{1}{n!} \left( \mathbf{r} \cdot \frac{\partial}{\partial \mathbf{R}} \right)^{n-1} \left( \frac{2}{n+1} \boldsymbol{\ell} + g_s \mathbf{s} \right) \cdot \mathbf{B}(\mathbf{R}, t).$$

## B.1.2  The Multipole Expansion

The total interaction energy between the atom and an external electromagnetic field is the sum of the interaction energies of all the individual atomic charges $q_\kappa$ at position $\mathbf{r}(\kappa)$ from $\mathbf{R}$. Note that we are not interested in the internal energy of the atom due to interactions between these particles. Using the results of Eqs. B.3 and B.8, the energy sums can be written in the form

$$\mathcal{V}_{\text{elec}}(t) = \sum_{\kappa} V_{\text{elec}}(\mathbf{r}_{\kappa}, t)$$

$$= -\sum_{i} \left[ \sum_{\kappa} q_{\kappa} r_i(\kappa) \right] E_i(\mathbf{R}, t)$$

$$- \frac{1}{2} \sum_{i} \sum_{j} \left[ \sum_{\kappa} q_{\kappa} r_i(\kappa) r_j(\kappa) \right] \frac{\partial}{\partial R_j} E_i(\mathbf{R}, t) + \dots$$

$$= \mathcal{V}_{E1}(t) + \mathcal{V}_{E2}(t) + \dots, \tag{B.9}$$

due to the electric field, and

$$\mathcal{V}_{\text{magn}}(t) = \sum_{\kappa} V_{\text{magn}}(\mathbf{r}_{\kappa}, t)$$

$$= -\sum_{i} \left[ \sum_{\kappa} \frac{q_{\kappa} \hbar}{2 m_{\kappa}} \left( \ell_i(\kappa) + g_s(\kappa) s_i(\kappa) \right) \right] B_i(\mathbf{R}, t)$$

$$- \sum_{i} \sum_{j} \left[ \sum_{\kappa} \frac{q_{\kappa} \hbar}{2 m_{\kappa}} \left( \frac{2}{3} \ell_i(\kappa) + g_s(\kappa) s_i(\kappa) \right) \right] \frac{\partial}{\partial R_j} B_i(\mathbf{R}, t) + \dots$$

$$= \mathcal{V}_{M1}(t) + \mathcal{V}_{M2}(t) + \dots, \tag{B.10}$$

due to the magnetic field, where each term $\mathcal{V}_{En}$ ($\mathcal{V}_{Mn}$) of the series can be factored into two parts; one containing all the atomic variables called the electric (magnetic) multipole moment, and the other expressing either the electric (magnetic) field or one of its spatial derivatives.

### B.1.3   The Dipole Approximation

A closer examination of the individual terms in the interaction energy sums given by Eqs. B.9 and B.10, reveals that $\mathcal{V}_{E1}(t)$ and $\mathcal{V}_{M1}(t)$ are by far the greatest contributing terms in the series. This is due to the fact that atoms range in size from approximately 10 pm up to around 300 pm (Slater 1964; Clementi et al. 1967), many orders of magnitude smaller than the wavelength of the visible light which is in the order of a few hundred nm. Hence for radiation within the optical range of the spectrum and above, where the wavelengths are much larger than the atomic dimensions, the contribution due to $\partial/\partial R$ and higher derivatives of the field becomes negligible and therefore the interaction is, to a good approximation, determined only by the first term in the series, referred to as the *dipole* term, which is directly proportional to the field strength.

Next, we turn our attention to the magnetic dipole term

$$\mathbf{m} = \sum_\kappa \frac{q_\kappa \hbar}{2m_\kappa} \left( \ell_i(\kappa) + g_s(\kappa)s_i(\kappa) \right). \tag{B.11}$$

Since protons and neutrons that constitute the atomic nuclei are some 2,000 times more massive than the electrons, the dominant contributions to the atomic magnetic moment comes from the electrons, i.e.

$$\mathbf{m} = -\frac{e\hbar}{2m_e}(\mathbf{L} + 2\mathbf{S}) = -\mu_B(\mathbf{L} + 2\mathbf{S}), \tag{B.12}$$

where $\mathbf{L}$ and $\mathbf{S}$ are the total electronic orbital and spin angular momenta respectively, $m_e$ is the electron mass,

$$\mu_B = -ea_0\frac{\alpha}{2} \tag{B.13}$$

is the Bohr magneton, $a_0 \simeq 0.53\,\text{Å}$ is the Bohr radius and $\alpha \simeq 1/137$ is the fine structure constant. The electric dipole on the other hand is given by

$$\mathbf{d} = \sum_\kappa q_\kappa \mathbf{r}(\kappa). \tag{B.14}$$

Since the atomic nucleus is nearly 1,000 times smaller than the atom itself, the product $q_\kappa \mathbf{r}(\kappa)$ is relatively negligible for the positively charged protons. Hence assuming a continuous electron charge distribution

$$\mathbf{d} = -e \int d^3 r \rho(\mathbf{r})\mathbf{r}, \tag{B.15}$$

where $\rho(\mathbf{r})$ is the charge density, i.e. the electron charge $e$ times the probability of finding an electron at position $\mathbf{r}$. In the case of a single electron in the $n$th bound state for example

$$\rho(\mathbf{r}) = -e\,|\psi_n(\mathbf{r})|^2 = -e\,\psi_n(\mathbf{r})^*\psi_n(\mathbf{r}), \tag{B.16}$$

where $\psi_n(\mathbf{r})$ is the electron wave-function for the $n$th state, and hence

$$\mathbf{d} = -e \int d^3 r \psi_n(\mathbf{r})^*\,\mathbf{r}\,\psi_n(\mathbf{r}) \equiv -e\langle n|\hat{\mathbf{r}}|n\rangle. \tag{B.17}$$

More generally it is possible to define a dipole *transition* moment

$$\mathbf{d}_{nm} \equiv \langle n|\hat{\mathbf{d}}|m\rangle = -e \int d^3 r \psi_n(\mathbf{r})^*\,\mathbf{r}\,\psi_m(\mathbf{r}) \equiv -e\langle n|\hat{\mathbf{r}}|m\rangle, \tag{B.18}$$

which describes the strength of the atomic transition from state $m$ to state $n$ due to interaction with the electromagnetic field. The ensemble of all possible transition moments between bound states forms a dipole transition matrix with $\mathbf{d}_{nm}$ defining the $n, m$ element of the matrix. Typically the values of $\mathbf{d}_{nm}$ are a fraction of $-ea_0$. Recalling the expression for the Bohr magneton in Eq. B.13, we find that in general the matrix elements of the magnetic dipole moments $\mathbf{m}$ tend to be smaller than those of the electric dipole moments $\mathbf{d}$ by at least a factor $\alpha/2$. This leads to the electric diploe approximation, whereby the interaction energy between an atom and an incident electromagnetic field is given by

$$|\mathcal{V}_{\text{int}}(t)| = -\mathbf{d} \cdot \mathbf{E}(\mathbf{R}, t). \tag{B.19}$$

In relation to the transition moments, it is instructive to note that in general a transition from state $n$ to $m$ is allowed if

1. The frequency of the incident electromagnetic radiation $f \approx \Delta E / h$, where $\Delta E$ is the energy difference between the two bound states, and
2. The interaction energy due to $\mathbf{d}_{nm}$ integral is non-zero.

In other words the radiation frequency is responsible for determining if an incoming photon is energetic enough to invoke a state transition, while $\mathbf{d}_{nm}$ is the probability that the photon would participate in this interaction.

## B.2   Single-Photon Interaction

### B.2.1   Rabi Model

*Main Reference (Shore 1990a)*

The Rabi model is a semi-classical model of a two-state atom, with $|1\rangle$ and $|2\rangle$ representing the ground and excited states respectively, interacting with an electromagnetic radiation field with frequency $\omega$ such that

$$\omega = \frac{1}{\hbar}(\mathcal{E}_2 - \mathcal{E}_1) + \Delta, \tag{B.20}$$

where $\mathcal{E}_2$ and $\mathcal{E}_1$ are the free atom energies associated with states $|1\rangle$ and $|2\rangle$, and $\Delta = \omega - \omega_0$ represents a detuning of the incoming radiation frequency $\omega$ from the atomic resonance frequency $\omega_0$. This interaction is then characterized by three simultaneous processes which can occur with varying probabilities (see Fig. B.2).

1. *Absorption:* the atom in state $|1\rangle$ may be excited to state $|2\rangle$ by absorbing a single photon from the incoming radiation field.

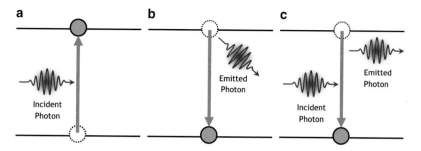

**Fig. B.2** Atom-field interaction in a two level system; (**a**) absorption, (**b**) spontaneous emission and (**c**) stimulated emission of photons

2. *Stimulated emission:* the atom in the excited state $|2\rangle$ is stimulated by a photon in the incoming radiation field (first predicted by Einstein (1917)) to undergo a transition back to state $|1\rangle$ and the resulting emitted photon will have the same optical mode (phase, frequency, etc.) as the incoming photon.
3. *Spontaneous emission:* the atom in the excited state $|2\rangle$ may, after some time, spontaneously decay into state $|1\rangle$ or any other lower energy states, emitting a single photon in a random optical mode.

In an ideal Rabi model, atomic states are assumed to be long lived with negligible spontaneous decay into other levels, and hence the first two processes dominate the atomic transitions. Nonetheless for situations where this assumption does not hold, spontaneous decay out of atomic states can be quantified via a decay rate $\Gamma$ and incorporated in the model dynamics (see Eq. B.42)

Starting with an initial wave-function for the state of the atom

$$|\Psi(t=0)\rangle = \alpha_1(0)|1\rangle + \alpha_2(0)|2\rangle, \tag{B.21}$$

the Rabi model provides a simple means for determining the time varying probabilities

$$P_1(t) \equiv |\alpha_1(t)|^2 \quad \text{and} \quad P_2(t) \equiv |\alpha_2(t)|^2, \tag{B.22}$$

for finding the atom in either states as a result of its continued interaction with the radiation field. The Schrödinger equation governing the time evolution of the atom-field system is given by

$$i\hbar \frac{\partial}{\partial t}|\Psi(t)\rangle = \hat{\mathcal{H}}|\Psi(t)\rangle, \tag{B.23}$$

or equivalently

$$i\hbar \frac{\partial}{\partial t}\begin{pmatrix} \alpha_1(t) \\ \alpha_2(t) \end{pmatrix} = \hat{\mathcal{H}}\begin{pmatrix} \alpha_1(t) \\ \alpha_2(t) \end{pmatrix}, \tag{B.24}$$

where the Hamiltonian

$$\hat{\mathcal{H}} = \hat{\mathcal{H}}_{\text{atom}} + \hat{\mathcal{H}}_{\text{int}}$$

$$= \begin{pmatrix} \mathcal{E}_1 & 0 \\ 0 & \mathcal{E}_2 \end{pmatrix} + \begin{pmatrix} \mathcal{V}_{11}(t) & \mathcal{V}_{12}(t) \\ \mathcal{V}_{21}(t) & \mathcal{V}_{22}(t) \end{pmatrix}, \tag{B.25}$$

and $\mathcal{V}_{ij}(t)$ are the atom-field interaction energies given by Eq. B.19. Evaluating Eq. B.18 reveals that for most atoms the diagonal elements of the electric dipole matrix $\mathbf{d}_{nn} = 0$ (i.e. atoms in an energy eigenstate do not possess an electric dipole moment) and $\mathbf{d}_{nm} = \mathbf{d}_{mn}^*$. Consequently $\mathcal{V}_{11}(t) = \mathcal{V}_{22}(t) = 0$ and

$$\hat{\mathcal{H}} = \begin{pmatrix} \mathcal{E}_1 & \mathcal{V}^*(t) \\ \mathcal{V}(t) & \mathcal{E}_2 \end{pmatrix} \tag{B.26}$$

where,

$$\mathcal{V}(t) = -\mathbf{d}_{21} \cdot \mathbf{E}(z, t) = \frac{\hbar}{2} \left( \Omega e^{-i\omega t} + \tilde{\Omega} e^{i\omega t} \right), \tag{B.27}$$

$$\hbar\Omega = -\mathbf{d}_{21} \cdot \mathbf{E}^{(+)}(z) \quad \text{and} \quad \hbar\tilde{\Omega} = -\mathbf{d}_{21} \cdot \mathbf{E}^{(-)}(z), \tag{B.28}$$

using a radiation field represented by Eq. A.12. The solution to Eq. B.23 involving the above time-dependant Hamiltonian is facilitated by utilizing a Rotating Wave reference frame.

### B.2.2   The Interaction Picture

*Main Reference (Shore 1990a)*

In quantum mechanics, the *interaction picture* or *Dirac picture* or *rotating wave picture* is an intermediate picture between the *Schrödinger picture* and the *Heisenberg picture*. Whereas in the other two pictures either the state vector or the operators carry time dependence, in the interaction picture both carry part of the time dependence of observables. More specifically, considering the Schrödinger equation in B.23, it is not necessary for the time-dependence of $\Psi(t)$ to be placed entirely within the unknown coefficients $\alpha(t)$. It is in fact often useful to assign some portion of the time-dependance to a phase $\phi(t)$ such that

$$|\Psi(t)\rangle = \Lambda|\overline{\Psi}(t)\rangle, \tag{B.29}$$

where $|\overline{\Psi}(t)\rangle = \overline{\alpha}_1(t)|1\rangle + \overline{\alpha}_2(t)|2\rangle$ and

$$\Lambda = \begin{pmatrix} e^{-i\phi_1(t)} & 0 \\ 0 & e^{-i\phi_2(t)} \end{pmatrix}. \tag{B.30}$$

By doing so we effectively transform the problem into a new reference frame in which we have introduced a time-dependant phase rotation $\phi_1(t) - \phi_2(t)$ between the components $\overline{\alpha}_1(t)$ and $\overline{\alpha}_2(t)$ while leaving the probabilities $P_1(t)$ and $P_2(t)$ in Eq. B.22 unaffected. Now rewriting the Schrödinger equation in this new rotating frame we obtain

$$i\hbar\frac{\partial}{\partial t}|\overline{\Psi}(t)\rangle = \underbrace{\left(\Lambda^\dagger\hat{\mathcal{H}}\Lambda - i\hbar\Lambda^\dagger\dot{\Lambda}\right)}_{\hat{\mathcal{H}}_I}|\overline{\Psi}(t)\rangle, \tag{B.31}$$

where

$$\hat{\mathcal{H}}_I = \begin{pmatrix} \mathcal{E}_1 - \hbar\dot{\phi}_1(t) & \mathcal{V}^*(t)e^{i\phi_1(t)-i\phi_2(t)} \\ \mathcal{V}(t)e^{i\phi_2(t)-i\phi_1(t)} & \mathcal{E}_2 - \hbar\dot{\phi}_2(t) \end{pmatrix} \tag{B.32}$$

is the Hamiltonian in the interaction picture.

Representing the problem in this reference frame is particularly useful since we are at liberty to assign arbitrary values to the phase $\phi(t)$ and its derivative $\dot{\phi}(t)$ in ways that can simplify the representation of the Hamiltonian. In particular, introducing the definitions

$$\phi_2(t) = \phi_1(t) + \omega t, \tag{B.33}$$

$$\hbar\dot{\phi}_1(t) = \mathcal{E}_1 + \frac{1}{2}\hbar(\omega_0 - \omega), \tag{B.34}$$

$$\hbar\dot{\phi}_2(t) = \mathcal{E}_1 + \frac{1}{2}\hbar(\omega_0 + \omega), \tag{B.35}$$

in conjunction with Eq. B.27, reduces the Hamiltonian to

$$\hat{\mathcal{H}}_I = \frac{\hbar}{2}\begin{pmatrix} \Delta & \Omega^* + \tilde{\Omega}^*e^{-2i\omega t} \\ \Omega + \tilde{\Omega}e^{2i\omega t} & -\Delta \end{pmatrix}. \tag{B.36}$$

## B.2.3 Rotating Wave Approximation

As we will see from the solution of the Schrödinger equation, the wave-function $|\overline{\Psi}(t)\rangle$ changes very little during an optical period $2\pi/\omega$ and the behavior of primary interest takes place over many optical cycles. On any appreciable time scale however the components $\tilde{\Omega}e^{2i\omega t}$ and $\tilde{\Omega}^*e^{-2i\omega t}$ undergo numerous oscillations which rapidly average to 0. The Rotating Wave Approximation (RWA) involves neglecting these high frequency terms which yields

$$\hat{\overline{\mathcal{H}}}_I = \frac{\hbar}{2}\begin{pmatrix} \Delta & \Omega^* \\ \Omega & -\Delta \end{pmatrix}. \tag{B.37}$$

It is also useful (see Appendix B.2.5) to introduce an equivalent representation of this Hamiltonian given by

$$\hat{\tilde{\mathcal{H}}}_I = \frac{\hbar}{2}\Delta \left(|1\rangle\langle 1| - |2\rangle\langle 2|\right) + \frac{\hbar}{2}\left(\Omega|1\rangle\langle 2| + \Omega^*|2\rangle\langle 1|\right)$$

$$\equiv \frac{\hbar}{2}\Delta\hat{\sigma}_z + \frac{\hbar}{2}\left(\Omega\hat{\sigma}_+ + \Omega^*\hat{\sigma}_-\right),\qquad\text{(B.38)}$$

where

$$\hat{\sigma}_z \equiv |1\rangle\langle 1| - |2\rangle\langle 2| \equiv \begin{pmatrix} 1 & 0 \\ 0 & -1 \end{pmatrix}\qquad\text{(B.39)}$$

is the Pauli $z$ matrix and $\hat{\sigma}_+ \equiv |2\rangle\langle 1|$ and $\hat{\sigma}_- \equiv |1\rangle\langle 2|$ are known as the raising and lowering operators.

An important advantage of working in the rotating wave picture is that with appropriate choice of $\dot{\phi}(t)$ the RWA Hamiltonian becomes time-independent and hence the Schrödinger equation B.31 assumes the closed form solution

$$|\overline{\Psi}(t)\rangle = \hat{\mathcal{U}}|\overline{\Psi}(0)\rangle,\qquad\text{(B.40)}$$

where the evolution matrix

$$\hat{\mathcal{U}}(t) = \exp\left(-\frac{i}{\hbar}\hat{\tilde{\mathcal{H}}}t\right)$$

$$= \begin{pmatrix} \cos\left(\frac{\Omega_R t}{2}\right) - i\frac{\Delta}{\Omega_R}\sin\left(\frac{\Omega_R t}{2}\right) & -i\frac{\Omega^*}{\Omega_R}\sin\left(\frac{\Omega_R t}{2}\right) \\ -i\frac{\Omega}{\Omega_R}\sin\left(\frac{\Omega_R t}{2}\right) & \cos\left(\frac{\Omega_R t}{2}\right) + i\frac{\Delta}{\Omega_R}\sin\left(\frac{\Omega_R t}{2}\right) \end{pmatrix}$$

$$\text{(B.41)}$$

represents a unitary transformation, $\Omega_R = \sqrt{|\Omega|^2 + \Delta^2}$ is the generalized Rabi frequency and the sinusoidal population transfer is referred to as *Rabi flopping*. More generally, one can also include the possibility of irreversible probability loss out of level $i$ at the rate $2\gamma_i = \Gamma_i$ by rewriting the RWA Hamiltonian as

$$\hat{\tilde{\mathcal{H}}}_I = \frac{\hbar}{2}\begin{pmatrix} \Delta & \Omega^* \\ \Omega & -\Delta \end{pmatrix} - \frac{\hbar}{2}\begin{pmatrix} i\Gamma_1 & 0 \\ 0 & i\Gamma_2 \end{pmatrix}.\qquad\text{(B.42)}$$

Figures B.3a and B.3c show Rabi flopping in the above RWA solution and the effect, on the quantum state evolution, of varying the detuning parameter and including loss terms in the Hamiltonian. A comparison is also drawn, in Figs. B.3b

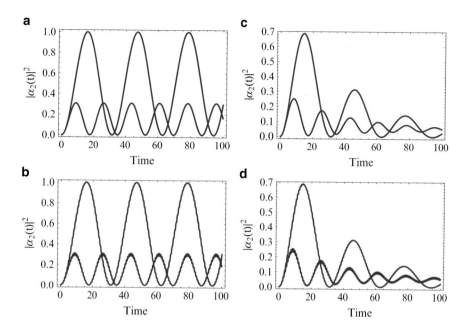

**Fig. B.3** The excited state of a two-level atom undergoing Rabi flopping with frequency $\Omega = 0.2$ and detuning $\Delta = 0$ (*blue*) and 0.3 (*red*). Plots (**a**) and (**c**) are derived using the RWA evolution operator in Eq. B.41, while (**b**) and (**d**) show the corresponding modeling using a direct numerical solution given by Eq. B.44 with $\omega = 20\Omega$. Decay rates in (**c**) and (**d**) are $\Gamma_1 = 0.02$ and $\Gamma_2 = 0.03$ (*blue*) as well as $\Gamma_1 = 0.05$ and $\Gamma_2 = 0$ (*red*)

and B.3d, between the RWA solution and a numerical solution derived directly from the Hamiltonian in Eq. B.26. Assuming a stationary atom whose center of mass coincides with the coordinate origin, $\mathcal{V}(t) = -\mathbf{d}_{21} \cdot \mathbf{E}(z = 0, t) = \hbar\Omega \cos(\omega t)$ and we may rewrite the time-dependant Hamiltonian (including decay rates) as

$$\hat{\mathcal{H}}(t) = \frac{\hbar}{2} \begin{pmatrix} -\omega_0 & 2\Omega^* \cos(\omega t) \\ 2\Omega \cos(\omega t) & \omega_0 \end{pmatrix} - \frac{\hbar}{2} \begin{pmatrix} i\Gamma_1 & 0 \\ 0 & i\Gamma_2 \end{pmatrix}, \tag{B.43}$$

where the resonant frequency $\omega_0 = \omega - \Delta$ and we have elected to fix the zero energy level at the mid-point between the ground and excited energy levels $\mathcal{E}_1$ and $\mathcal{E}_2$. The operator $\hat{\mathcal{U}}(t)$ can then be expressed as a series of infinitesimally short time-independent evolutionary stages such that

$$\hat{\mathcal{U}}(t) \approx \exp\left(-\frac{i}{\hbar}\hat{\mathcal{H}}(t_N)\delta t\right) \ldots \exp\left(-\frac{i}{\hbar}\hat{\mathcal{H}}(t_0)\delta t\right), \tag{B.44}$$

where $t_i = i\delta t$ and $\delta t \ll 1/\omega$.

## B.2.4   Separable Hamiltonians

Since our choice of the rotating phase matrix $\Lambda$ is somewhat arbitrary, it is possible to adopt a more systematic approach for systems whose Hamiltonian in the Shrödinger picture can be written in the form

$$\hat{\mathcal{H}} = \hat{\mathcal{H}}_0 + \hat{\mathcal{H}}_{\text{int}}, \tag{B.45}$$

where, for a composite system, $\hat{\mathcal{H}}_0$ often incorporates the simple and exactly solvable Hamiltonians of all the individual components in isolation, while the interaction Hamiltonian $\hat{\mathcal{H}}_{\text{int}}$ captures the physics governing the interactions between the various components within the system. In this commonly occurring scenario, letting

$$\Lambda = \exp\left(-\frac{i}{\hbar}\hat{\mathcal{H}}_0 t\right) \tag{B.46}$$

allows a simplification of Eq. B.31 such that the Hamiltonian in the interaction picture is

$$\hat{\mathcal{H}}_I = \Lambda^\dagger\left(\hat{\mathcal{H}}_0 + \hat{\mathcal{H}}_{\text{int}}\right)\Lambda - \Lambda^\dagger\left(\hat{\mathcal{H}}_0\Lambda\right)$$
$$= \Lambda^\dagger\hat{\mathcal{H}}_{\text{int}}\Lambda. \tag{B.47}$$

Applying this to the Rabi Hamiltonian in Eq. B.25 with

$$\hat{\mathcal{H}}_0 = \frac{\hbar}{2}\begin{pmatrix} -\omega_0 & 0 \\ 0 & \omega_0 \end{pmatrix}, \tag{B.48}$$

and making the rotating wave approximation we find

$$\hat{\mathcal{H}}_I = \frac{\hbar}{2}\begin{pmatrix} 0 & \Omega^* e^{i\Delta t} \\ \Omega e^{-i\Delta t} & 0 \end{pmatrix}. \tag{B.49}$$

Note that the difference between this representation and the RWA Hamiltonian in Eq. B.37 is only due to our choice of the rotating wave reference frame. Whereas Eq. B.37 represents a reference frame that is rotating with the angular speed $\dot{\phi}_2(t) - \dot{\phi}_1(t) = \omega$, here we have elected to use a rotating wave frame with the angular speed $\dot{\phi}_2(t) - \dot{\phi}_1(t) = \mathcal{E}_2/\hbar - \mathcal{E}_1/\hbar = \omega_0$. Although the choice of $\Lambda$ is entirely problem dependant, using the formalism of Eq. B.47 systematizes the extension of the interaction picture to arbitrary systems with large Hilbert spaces.

## B.2.5   Jaynes-Cummings Model

*Main References (Shore and Knight 1993; Cives-esclop and Sánchez-Soto 1999)*

This model is a fully quantum electrodynamic version of the Rabi model, where the classical electric field is replaced by its quantized counterpart and the Schrödinger equation determines the time evolution of the *composite* atom-field quantum state. Similar to the Rabi model, the two-level atom is represented by

$$|\psi(t)\rangle_{\text{atom}} = \alpha_1(t)|1\rangle + \alpha_2(t)|2\rangle, \tag{B.50}$$

while the state of the quantum field is given by the generalized form in Eq. A.38. Taking, for simplicity, the Fock states of a single-mode light

$$|\psi(t)\rangle_{\text{field}} = \sum_{n=0}^{\infty} \beta_n(t)|n\rangle, \tag{B.51}$$

the nonseparable atom-field composite state is then given by the wave-function

$$|\Psi(t)\rangle = \sum_{n=0}^{\infty} \sum_{m=1}^{2} c_{n,m}(t)|n,m\rangle, \tag{B.52}$$

where $c_{n,m}$ are arbitrary complex coefficients. Let us now consider the atom initially in state $|1\rangle$ and the field initially in state $|n+1\rangle$. The radiation field makes a transition to state $|n\rangle$, loosing one photon and exciting the atom to state $|2\rangle$ in the process, and vice versa. Hence the two composite states

$$|n+1,1\rangle \equiv |\Psi_1\rangle \text{ and } |n,2\rangle \equiv |\Psi_2\rangle$$

are naturally connected via a single photon interaction and we may re-introduce a two-level Schrödinger equation (Eq. B.24)

$$i\hbar \frac{\partial}{\partial t} \begin{pmatrix} c_{n,1}(t) \\ c_{n,2}(t) \end{pmatrix} = \hat{\mathcal{H}}_{\text{JC}} \begin{pmatrix} c_{n,1}(t) \\ c_{n,2}(t) \end{pmatrix}. \tag{B.53}$$

The complete Hamiltonian governing the Jaynes-Cummings model extends Eq. B.25 by including an additional term associated with the quantized field energy, now part of the evolutionary dynamics. Hence

$$\hat{\mathcal{H}}_{\text{JC}} = \hat{\mathcal{H}}_{\text{field}} + \hat{\mathcal{H}}_{\text{atom}} + \hat{\mathcal{H}}_{\text{int}}, \tag{B.54}$$

where the additional term

$$\hat{\mathcal{H}}_{\text{field}} = \hbar\omega\left(\hat{a}\hat{a}^{\dagger} + \frac{1}{2}\right) \tag{B.55}$$

is given by Eq. A.32 for a single mode field. As before

$$\hat{\mathcal{H}}_{\text{atom}} = \begin{pmatrix} \mathcal{E}_1 & 0 \\ 0 & \mathcal{E}_2 \end{pmatrix} = \frac{1}{2}(\mathcal{E}_1 + \mathcal{E}_2)I - \frac{\hbar\omega_0}{2}\hat{\sigma}_z, \tag{B.56}$$

where $\mathcal{E}_1$ and $\mathcal{E}_1$ are the energy eigenvalues for the non-interacting atom, $I$ is the unitary matrix and $\hat{\sigma}_z \equiv |1\rangle\langle1| - |2\rangle\langle2|$ is the Pauli $z$ matrix, and

$$\hat{\mathcal{H}}_{\text{int}} = \begin{pmatrix} 0 & \mathcal{V}(t)^* \\ \mathcal{V}(t) & 0 \end{pmatrix} = \mathcal{V}(t)\hat{\sigma}_+ + \mathcal{V}(t)^*\hat{\sigma}_-, \tag{B.57}$$

where $\hat{\sigma}_+ = |2\rangle\langle1|$ and $\hat{\sigma}_- = |1\rangle\langle2|$ represent the raising and lowering operators, $\mathcal{V}(t) = -\mathbf{d}_{21} \cdot \hat{\mathbf{E}}(z,t)$ is the interaction energy, and $\mathbf{d}_{21}$ and $\hat{\mathbf{E}}(z,t)$ represent the dipole transition matrix and the electric field operator respectively. Now dropping the unimportant global phase from Eqs. B.55 and B.56 leads to the standard Jaynes-Cummings Hamiltonian

$$\hat{\mathcal{H}}_{\text{JC}} = \hbar\omega\hat{a}^{\dagger}\hat{a} - \frac{\hbar}{2}\omega_0\hat{\sigma}_z + \frac{\hbar}{2}\left(\mathcal{V}\hat{\sigma}_+ + \mathcal{V}^*\hat{\sigma}_-\right).$$

A system commonly considered in the study of the Jaynes-Cummings model is a two level atom inside an optical cavity where, in contrast to the traveling wave electric field represented by Eq. A.34, the electric field operator is derived from the *standing wave* solution of the wave equation in Eq. A.3. Consequently the operator for a single-mode field is given by

$$\hat{\mathbf{E}} = i\sqrt{\frac{\hbar\omega}{2\epsilon_0 V}}\mathbf{u}(\hat{a} + \hat{a}^{\dagger})\sin(kz + \phi), \tag{B.58}$$

where $\mathbf{u}$ is the polarized unit vector, $\omega$ is the radiation frequency, $k$ is the wave number, $V$ is the normalized volume of the radiation field, $z$ denotes the distance to the center of the trap, and $\hat{a} \equiv \hat{a}(t) \equiv \hat{a}(0)e^{-i\omega t}$ and $\hat{a}^{\dagger} \equiv \hat{a}^{\dagger}(t) \equiv \hat{a}^{\dagger}(0)e^{i\omega t}$ are the annihilation and creation operators of the single mode field respectively, with an implicit time component (Luo et al. 1998). Using $\hat{\mathcal{H}}_0 = \hat{\mathcal{H}}_{\text{field}} + \hat{\mathcal{H}}_{\text{atom}}$ we may now find the Hamiltonian in the interaction picture (see Appendix B.2.4) and apply the rotating wave approximation

$$\hat{\bar{\mathcal{H}}}_{\text{JC},I} = \hbar g_0(\hat{a}(0)e^{-i\Delta t}\hat{\sigma}_+ + \hat{a}^{\dagger}(0)e^{i\Delta t}\hat{\sigma}_-), \tag{B.59}$$

where we have assumed a real valued dipole transition $\mathbf{d}$,

$$g_0 = i \sqrt{\frac{\omega}{2\hbar\epsilon_0 V}} \sin(kz + \phi) \mathbf{u} \cdot \mathbf{d} \tag{B.60}$$

is the atom-field coupling and $\Delta = \omega - \omega_0$ is the detuning from the resonance frequency. To obtain the Jaynes-Cummings Hamiltonian in its most common form we use the inverse of Eq. B.47 to convert the RWA Hamiltonian $\hat{\bar{\mathcal{H}}}_{JC,I}$ from the interaction picture back to the Schrödinger picture, whereby we find

$$\hat{\bar{\mathcal{H}}}_{JC} = \hat{\mathcal{H}}_0 + \hat{\bar{\mathcal{H}}}_{int}$$

$$= \hbar\omega\hat{a}^\dagger\hat{a} - \frac{\hbar}{2}\omega_0\hat{\sigma}_z + \hbar g_0(\hat{a}\hat{\sigma}_+ + \hat{a}^\dagger\hat{\sigma}_-). \tag{B.61}$$

It is now possible to determine the action of the RWA Hamiltonian operator on individual states of the composite system, where

$$\hat{\bar{\mathcal{H}}}_{JC}|\Psi_1\rangle = \hbar\omega(n+1)|\Psi_1\rangle - \frac{\hbar}{2}\omega_0|\Psi_1\rangle + \hbar g_0\sqrt{n+1}|\Psi_2\rangle, \text{ and} \tag{B.62}$$

$$\hat{\bar{\mathcal{H}}}_{JC}|\Psi_2\rangle = \hbar\omega n|\Psi_2\rangle + \frac{\hbar}{2}\omega_0|\Psi_2\rangle + \hbar g_0\sqrt{n+1}|\Psi_1\rangle, \tag{B.63}$$

and hence rewriting the Hamiltonian in its matrix form

$$\hat{\bar{\mathcal{H}}}_{JC}^{(n)} = \hbar \begin{pmatrix} \omega(n+1) - \omega_0/2 & g_0\sqrt{n+1} \\ g_0\sqrt{n+1} & \omega n + \omega_0/2 \end{pmatrix}. \tag{B.64}$$

As with the Rabi model, the Schrödinger equation B.53 also finds an exact closed form analytical solution given by

$$\hat{\mathcal{U}}_{JC}^{(n)} = \exp\left(-\frac{i}{\hbar}\hat{\mathcal{H}}_{JC}^{(n)}t\right), \tag{B.65}$$

which for the simple resonant case where $\Delta = 0$ yields

$$\hat{\mathcal{U}}_{JC}^{(n)} = e^{-\frac{1}{2}i(2n+1)t\omega} \begin{pmatrix} \cos(\Omega_R t) & -i\sin(\Omega_R t) \\ -i\sin(\Omega_R t) & \cos(\Omega_R t) \end{pmatrix}, \tag{B.66}$$

where

$$\Omega_R = g_0\sqrt{n+1} \tag{B.67}$$

is the quantum electrodynamic Rabi frequency. Notably, in the fully quantum model, there are Rabi oscillations even for the case when $n = 0$ known as vacuum-field Rabi oscillations. Two-photon generalizations of the Jaynes-Cummings Model can be found in Wódkiewicz and Eberly (1985) and Gerry (1985).

## B.2.6   The a.c. Stark Effect or Light Shift

*Main References (Foot 2005; Metcalf and der Straten 1999a; Nakahara and Ohmi 2008)*

The a.c. Stark effect refers to the shift in the atomic energy eigenstates when interacting with an external periodic or a.c. electric field. This is in contrast to the static or d.c. Stark effect where the energy shift is due to a uniform field.

Considering a Hamiltonian of the form given by Eq. B.45, the presence of off-diagonal interaction terms in $\hat{\mathcal{H}}_I$ dictates that the diagonal elements $\mathcal{E}_1$ and $\mathcal{E}_2$ of the non-interacting Hamiltonian $\hat{\mathcal{H}}_0$ no longer represent the energy eigenstates of the total Hamiltonian. In the rotating wave picture, the energy eigenvalues of the RWA Hamiltonian in Eq. B.37 are determined by solving the eigenvalue equation

$$\begin{vmatrix} \Delta/2 - \lambda & \Omega^*/2 \\ \Omega/2 & -\Delta/2 - \lambda \end{vmatrix} = \lambda^2 - \left(\frac{\Delta}{2}\right)^2 - \left(\frac{|\Omega|}{2}\right)^2 = 0, \qquad (B.68)$$

which yields

$$\lambda = \pm\frac{1}{2}\sqrt{\Delta^2 + |\Omega|^2} = \pm\left(\frac{\Delta}{2} + \Omega'\right), \qquad (B.69)$$

where the new energy eigenvalues $\overline{\mathcal{E}} = \hbar\lambda$ with $\Omega'$ representing additional terms in the Taylor expansion. In the absence of an external light field $\Omega = \Omega' = 0$ and the unperturbed eigenvalues are $\lambda = \pm\Delta/2$ corresponding to the unperturbed states $|1\rangle$ and $|2\rangle$. As depicted in Fig. B.4 however, interaction with an external light field introduces a shift $\Delta\mathcal{E} = \pm\hbar\Omega'$ in the atomic energies and the new eigenstates of the system are known as *dressed states*. Normally energy shifts are most important at large frequency detuning, where $|\Delta| \gg \Omega$ and hence the effect of absorption is negligible. In this case the energy eigenvalues are given by

$$\lambda \approx \pm\left(\frac{\Delta}{2} + \frac{|\Omega|^2}{4\Delta}\right). \qquad (B.70)$$

A similar approach may be used when it becomes necessary to take into account the spontaneous emission process in which case the RWA Hamiltonian is given by Eq. B.42 and we assume a stable ground state with $\Gamma_1 = 0$ and a leaky excited state with $\Gamma_2 = \Gamma > 0$. Consequently the new eigenvalues are given by

$$\lambda = -\frac{i\Gamma}{4} \pm \sqrt{\left(\frac{\Delta}{2} + \frac{i\Gamma}{4}\right)^2 + \frac{|\Omega|^2}{2}}$$

$$= \pm\left(\frac{\Delta}{2} + \Omega'\right), \qquad (B.71)$$

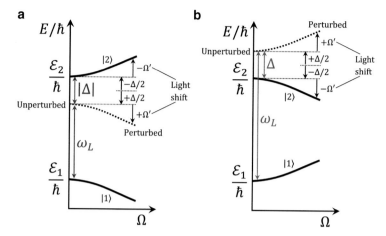

**Fig. B.4** Eigenenergies of a two-level atom interacting with an external light field producing an a.c. Stark effect for (**a**) negative (*red*) and (**b**) positive (*blue*) detuning, as a function of Rabi frequency

where for a first order expansion

$$\Omega' \approx \frac{|\Omega|^2}{4\,(\Delta + i\,\Gamma/2)}, \tag{B.72}$$

the imaginary part of which

$$\text{Im}\,(\Omega') = \frac{\Gamma\,|\Omega|^2}{8\,(\Delta^2 + \Gamma^2/4)}, \tag{B.73}$$

is associated with the loss due to spontaneous photon scattering, while its real part

$$\text{Re}\,(\Omega') = \frac{\Delta\,|\Omega|^2}{4\,(\Delta^2 + \Gamma^2/4)}, \tag{B.74}$$

determines the energy shift $\Delta\mathcal{E} = \pm\hbar\,\text{Re}\,(\Omega')$ due to the atom-field interaction. The dependance of energy shift on the sign of $\Delta$ has important consequences for dipole-force trapping of atoms, as described in Appendix E.2.2.

## B.3 Two-Photon Interaction

### B.3.1 Raman Process

*Main Reference (Shore 1990c)*

The Raman process, Raman effect or Raman scattering is a semi-classical model for a three-state system in a $\Lambda$ formation, whereby transitions between an initial state

**Fig. B.5** Schematic diagram
of the Raman process

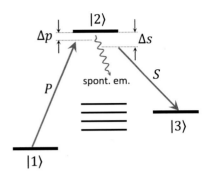

$|1\rangle$ and a final state $|3\rangle$ occur via a two-photon process involving an intermediary
state $|2\rangle$ (see Fig. B.5). During the transition $|1\rangle \longrightarrow |2\rangle$ the atom absorbs a single
photon from a *pump radiation* field with frequency

$$\omega_1 = \frac{1}{\hbar}(\mathcal{E}_2 - \mathcal{E}_1) + \Delta_P, \qquad (B.75)$$

and detuning $\Delta_P$. Naturally, the atom in state $|2\rangle$ may, after some time, sponta-
neously decay into a lower energy level other than $|1\rangle$, emitting a photon in a
random optical mode. This is referred to as *spontaneous Raman scattering* which is
typically a very weak process and is used in Raman spectroscopy. A *stimulated
Raman process*, on the other hand, involves employing an additional radiation
source, known as *dump radiation*, with frequency

$$\omega_2 = \frac{1}{\hbar}(\mathcal{E}_2 - \mathcal{E}_3) + \Delta_S, \qquad (B.76)$$

and detuning $\Delta_S$, to stimulate and amplify a predetermined transition $|2\rangle \longrightarrow |3\rangle$.

If $\mathcal{E}_3 > \mathcal{E}_1$, the process leads to an overall increase in the atomic energy and the
emitted radiation is referred to as *Stokes radiation*. On the other hand if $\mathcal{E}_3 < \mathcal{E}_1$, the
process leads to an overall decrease in the atomic energy and the emitted radiation is
referred to as *anti-Stokes radiation*. The case where $\mathcal{E}_3 = \mathcal{E}_1$ is known as *Rayleigh
scattering* in which there is no energy exchange between photons and the atom.

The Raman process can be further classified as non-resonance and resonance
Raman, depicted in Fig. B.6. In the non-resonance Raman effect, the intermediate
state $|2\rangle$ is a *virtual* energy level (i.e. not an eigenstate) and the incoming laser is far
detuned from electronic transitions of the atom. In resonance Raman spectroscopy
on the other hand, the energy of the incoming laser is adjusted such that it coincides
with an electronic transition of the atom, resulting in a greatly increased scattering
intensity. The Raman process differs from the process of fluorescence in which the
incident light is completely absorbed and the system is transferred to an excited state
before decaying into various lower states only after a certain resonance lifetime. In
contrast, the Raman effect is a coherent process which occurs over a much shorter
time-scale and can take place for any frequency of the incident light.

**Fig. B.6** Electronic transitions in three distinct two-photon $\Lambda$ processes: non-resonance Raman, resonance Raman and fluorescence

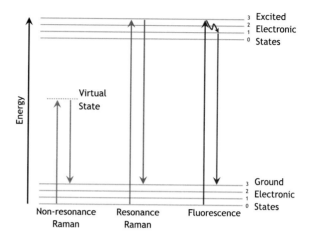

The generalized three-state RWA equations can be obtained by extending Eq. B.29, where

$$\Lambda = \begin{pmatrix} e^{-i\phi_1(t)} & 0 & \\ 0 & e^{-i\phi_2(t)} & 0 \\ 0 & 0 & e^{-i\phi_3(t)} \end{pmatrix}. \tag{B.77}$$

With the appropriate phase choices given by

$$\phi_1(t) = \phi_2(t) + \omega_1 t, \tag{B.78}$$

$$\phi_3(t) = \phi_2(t) + \omega_2 t, \tag{B.79}$$

the three-state RWA Hamiltonian then becomes

$$\hat{\mathcal{H}} = \frac{\hbar}{2} \begin{pmatrix} 2\Delta_1 - i\Gamma_1 & \Omega_1^* & 0 \\ \Omega_1 & 2\Delta_2 - i\Gamma_2 & \Omega_2^* \\ 0 & \Omega_2 & 2\Delta_3 - i\Gamma_3 \end{pmatrix}, \tag{B.80}$$

where $\hbar\Delta_i = \mathcal{E}_i - \hbar\dot{\phi}_i(t)$ for level $i = 1, 2, 3$, $\Gamma_i$ represents the rate of spontaneous decay out of level $i$ and $\hbar\Omega_1 \equiv \hbar\Omega_P = -\mathbf{d}_{21} \cdot \mathbf{E}_P$ and $\hbar\Omega_2 \equiv \hbar\Omega_S = -\mathbf{d}_{32} \cdot \mathbf{E}_S$ are the interaction energies associated with pump and Stokes radiation fields $\mathbf{E}_P$ and $\mathbf{E}_S$ with respective frequencies $\omega_1$ and $\omega_2$.

In the case of states in a $\Lambda$ formation we may choose

$$\dot{\phi}_2(t) = \mathcal{E}_2(t)/\hbar, \tag{B.81}$$

$$\dot{\phi}_1(t) = \dot{\phi}_2(t) - \omega_1, \tag{B.82}$$

$$\dot{\phi}_3(t) = \dot{\phi}_2(t) - \omega_2, \tag{B.83}$$

which gives

$$\Delta_1 = \hbar\omega_1 - (\mathcal{E}_2 - \mathcal{E}_1) = \Delta_P, \tag{B.84}$$

$$\Delta_2 = 0, \tag{B.85}$$

$$\Delta_3 = \hbar\omega_2 - (\mathcal{E}_2 - \mathcal{E}_3) = \Delta_S. \tag{B.86}$$

and the RWA Hamiltonian for the Raman process is reduced to

$$\hat{\bar{\mathcal{H}}} = \frac{\hbar}{2}\begin{pmatrix} 2\Delta_P - i\Gamma_1 & \Omega_P^* & 0 \\ \Omega_P & -i\Gamma_2 & \Omega_S^* \\ 0 & \Omega_S & 2\Delta_S - i\Gamma_3 \end{pmatrix}. \tag{B.87}$$

## B.3.2   Adiabatic Elimination

*Main Reference (Brion et al. 2007)*

Considering the non-resonance Raman effect where $|\Delta_{S,P}| \gg |\Omega_{1,2}|$, if the system is initially prepared in a superposition $\alpha_1|1\rangle + \alpha_3|3\rangle$, the virtual state $|2\rangle$ will then essentially remain unpopulated throughout the entire process, while second-order transitions will take place between the two lower states $|1\rangle$ and $|3\rangle$. The effective dynamics of such a system can then be approximately described by a $2 \times 2$ Hamiltonian after eliminating the irrelevant virtual state using a procedure known as *adiabatic elimination*.

In its simplified form, the procedure involves replacing Eq. B.81 with the choice of

$$\dot{\phi}_2(t) = \mathcal{E}_2(t)/\hbar + \Delta/2, \tag{B.88}$$

thus arriving at an alternative representation of the RWA Hamiltonian (not including the decay terms) given by

$$\hat{\bar{\mathcal{H}}} = \frac{\hbar}{2}\begin{pmatrix} \delta & \Omega_P^* & 0 \\ \Omega_P & -\Delta & \Omega_S^* \\ 0 & \Omega_S & -\delta \end{pmatrix}, \tag{B.89}$$

where $\Delta = \Delta_P + \Delta_S$ and $\delta = \Delta_P - \Delta_S$. Hence the Schrödinger equation

$$i\hbar\frac{\partial}{\partial t}\begin{pmatrix} \alpha_1 \\ \alpha_2 \\ \alpha_3 \end{pmatrix} = \hat{\bar{\mathcal{H}}}\begin{pmatrix} \alpha_1 \\ \alpha_2 \\ \alpha_3 \end{pmatrix}, \tag{B.90}$$

can be expressed as a set of differential equations

$$\begin{cases} i\hbar\dot{\alpha}_1 = \Omega_P^*\alpha_2 + \delta\alpha_1 \\ i\hbar\dot{\alpha}_2 = \Omega_P\alpha_1 - \Delta\alpha_2 + \Omega_S^*\alpha_3 \\ i\hbar\dot{\alpha}_3 = \Omega_S\alpha_2 - \delta\alpha_3 \end{cases} . \tag{B.91}$$

Assuming that the virtual state $|2\rangle$ remains unpopulated throughout the evolution we may set $\dot{\alpha}_2 = 0$ in the second equation, whereby

$$\alpha_2 = \frac{\Omega_S^*\alpha_3 + \Omega_P\alpha_1}{\Delta}. \tag{B.92}$$

Substituting this into the remaining two equations yields the effective Hamiltonian

$$\hat{\tilde{\mathcal{H}}}_{\text{eff}} = \frac{\hbar}{2\Delta} \begin{pmatrix} |\Omega_P|^2 + \delta\Delta & \Omega_T^* \\ \Omega_T & |\Omega_S|^2 - \delta\Delta \end{pmatrix}, \tag{B.93}$$

where $\Omega_T = \Omega_P\Omega_S$.

### B.3.3 Stimulated Raman Adiabatic Passage

*Main References (Bergmann et al. 1998; Gaubatz et al. 1990)*

Stimulated Raman Adiabatic Passage (STIRAP) is a variant of the stimulated Raman process in which overlapping pump and Stokes pulses with time varying amplitudes are employed to produce *complete* population transfer between two quantum states of an atom or molecule.

For the simplest implementation of STIRAP where the interaction energies are all real valued, and assuming long lived states $|1\rangle$ and $|3\rangle$ and a leaky state $|2\rangle$ with decay rate $\Gamma$, the time dependant RWA Hamiltonian reads

$$\hat{\tilde{\mathcal{H}}}(t) = \frac{\hbar}{2} \begin{pmatrix} 2\Delta_P & \Omega_P(t) & 0 \\ \Omega_P(t) & -i\Gamma & \Omega_S(t) \\ 0 & \Omega_S(t) & 2\Delta_S \end{pmatrix}, \tag{B.94}$$

where $\Delta_P$ and $\Delta_S$ denote the pump and Stokes radiation detuning, $\hbar\Omega_P(t) = -\mathbf{d}_{21} \cdot \mathbf{E}_P(t)$, $\hbar\Omega_S(t) = -\mathbf{d}_{32} \cdot \mathbf{E}_S(t)$, and $\mathbf{E}_P(t)$ and $\mathbf{E}_S(t)$ represent the time dependant electric field associated with pump and Stokes radiations.

For any Hamiltonian the eigenstates of the system define a set of states whose quantum amplitude remains unchanged, up to a phase factor, during the time evolution of the system. For the STIRAP Hamiltonain, these instantaneous eigenstates, also referred to as *dressed states*, are given by

$$|a^+\rangle = \sin\Theta \sin\Phi|1\rangle + \cos\Phi|2\rangle + \cos\Theta \sin\Phi|3\rangle, \qquad (B.95)$$

$$|a^0\rangle = \cos\Theta|1\rangle - \sin\Theta|3\rangle, \qquad (B.96)$$

$$|a^-\rangle = \sin\Theta \cos\Phi|1\rangle - \sin\Phi|2\rangle + \cos\Theta \sin\Phi|3\rangle, \qquad (B.97)$$

where the mixing angle $\Theta$ obeys

$$\tan\Theta(t) = \frac{\Omega_P(t)}{\Omega_S(t)}, \qquad (B.98)$$

and

$$\tan\Phi(t) = \frac{\sqrt{\Omega_P^2(t) + \Omega_S^2(t)}}{\sqrt{\Omega_P^2(t) + \Omega_S^2(t) + \Delta_P^2} - \Delta_P}. \qquad (B.99)$$

Crucially, the dressed state $|a^0\rangle$, also referred to as *dark state* or *trapped state*, lies in a plane perpendicular to $|2\rangle$. Hence if the system could be initially prepared and then maintained in this trapped state, it can never acquire a component of the leaky state $|2\rangle$. This means that the pump laser cannot transfer population from state $|1\rangle$ to the decaying intermediate state and instead the population is directly channeled into state $|3\rangle$. In practice, to ensure that the system persists in state $|a^0\rangle$, pulses with characteristic envelopes depicted in Fig. B.7a are applied in a counter-intuitive order. Rather than exposing the atom to the pump laser first and then to the Stokes laser, one reverses the ordering of the pulse sequence. It is easy to verify for this setup that at $t = 0$ we have $\Theta(0) = 0$ corresponding to trapped state $|a^0\rangle = |1\rangle$, and at $t = \infty$ we have $\Theta(\infty) = \pi/2$ corresponding to trapped state $|a^0\rangle = -|3\rangle$. For the system dynamics to follow the rotation of vector $|a^0\rangle$ adiabatically, the coupling of the states by the radiation fields needs to be sufficiently strong. For laser pulses with a smooth shape, a rule of thumb condition is given by

$$\Omega_{eff}\Delta\tau > 10, \qquad (B.100)$$

where $\Omega_{eff} = \sqrt{\Omega_P^2 + \Omega_S^2}$ and $\Delta\tau$ is the period during which the pulses overlap.

Since the STIRAP Hamiltonian is time-dependant with arbitrary elements $\Omega_P(t)$ and $\Omega_S(t)$, a closed form analytical solution for the Schrödinger equation does not generally exist. Instead Fig. B.7c depicts the time-evolution of all three states $|1\rangle$, $|2\rangle$ and $|3\rangle$ using a numerical method, whereby we have approximated the overall time-dependant evolution as a series of time-independent evolutionary stages. Hence the evolution matrix

$$\hat{U}(t) \approx \exp\left(-\frac{i}{\hbar}\hat{\mathcal{H}}_N \delta t\right)\ldots\exp\left(-\frac{i}{\hbar}\hat{\mathcal{H}}_0 \delta t\right), \qquad (B.101)$$

**Fig. B.7** Time evolution of
(**a**) the Rabi frequencies of
the pump and Stokes laser
(Gaussian envelopes with
$\sigma = 10$, peak amplitude 0.5
and peak to peak time interval
13.7), (**b**) the mixing angle
(see Eq. B.98); and (**c**)
Complete population transfer
from ground state $|1\rangle$ to
excited state $|3\rangle$ with minimal
participation of the
intermediary state $|2\rangle$ which
can, in principle be made
negligible. Were $|2\rangle$ to be a
leaky state with $\Gamma = 0.04$,
the impact on the population
transfer would still be small
(dotted curve) compared to
the substantial decay if
state $|2\rangle$ was fully
participating in the process

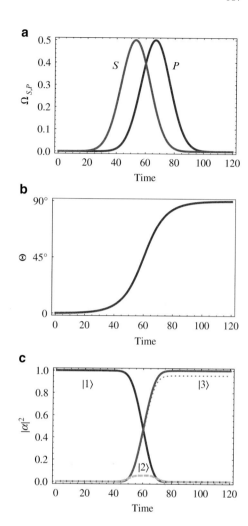

where

$$\hat{\overline{\mathcal{H}}}_k = \frac{\hbar}{2} \begin{pmatrix} 2\Delta_P & \Omega_P(k\delta t) & 0 \\ \Omega_P(k\delta t) & -i\Gamma & \Omega_S(k\delta t) \\ 0 & \Omega_S(k\delta t) & 2\Delta_S \end{pmatrix}, \tag{B.102}$$

$k = 0, 1, 2 \ldots N$, $\delta t = t/N$ and the integer $N$ can be made arbitrarily large to achieve the desired level of accuracy.

# Appendix C
# Bloch Rotations

*Main References (Benenti et al. 2004; Fox 2007)*

## C.1 Theory

In quantum mechanics, the Bloch sphere is a geometrical representation of the pure state space of a qubit or two-level quantum mechanical system depicted in Fig. C.1. In this representation, a two-level quantum system can, in its most general form, be written as

$$|\psi\rangle = \cos(\theta/2)|1\rangle + e^{i\varphi} \sin(\theta/2)|2\rangle, \tag{C.1}$$

where $0 < \theta < \pi$ and $0 < \varphi < 2\pi$ corresponding to Cartesian coordinates of a unit vector

$$\begin{cases} x = \sin(\theta)\cos(\varphi) \\ y = \sin(\theta)\sin(\varphi) \\ z = \cos(\theta) \end{cases} . \tag{C.2}$$

An important class of unitary transformations is *rotations* of the Bloch sphere about an arbitrary axis directed along unitary vector **n**. It is possible to show that the corresponding rotation operator takes the form

$$\begin{aligned} \hat{R}_{\mathbf{n}}(\delta) &= \exp\left(-i\frac{\delta}{2}(\mathbf{n}\cdot\vec{\sigma})\right) \\ &= \cos\left(\frac{\delta}{2}\right)I - i\sin\left(\frac{\delta}{2}\right)\mathbf{n}\cdot\vec{\sigma} \\ &= \cos\left(\frac{\delta}{2}\right)I - i\sin\left(\frac{\delta}{2}\right)\left(n_x\sigma_x + n_y\sigma_y + n_z\sigma_z\right), \end{aligned} \tag{C.3}$$

J. Wang and K. Manouchehri, *Physical Implementation of Quantum Walks*, Quantum Science and Technology, DOI 10.1007/978-3-642-36014-5, © Springer-Verlag Berlin Heidelberg 2014

**Fig. C.1** Schematic diagram
of a Bloch sphere

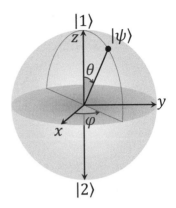

where $I$ is the $2 \times 2$ identity matrix and

$$\sigma_x = \begin{pmatrix} 0 & 1 \\ 1 & 0 \end{pmatrix}, \tag{C.4}$$

$$\sigma_y = \begin{pmatrix} 0 & -i \\ i & 0 \end{pmatrix}, \tag{C.5}$$

$$\sigma_z = \begin{pmatrix} 1 & 0 \\ 0 & -1 \end{pmatrix}, \tag{C.6}$$

are Pauli matrices satisfying the condition $\sigma_x^2 = \sigma_y^2 = \sigma_z^2 = I$. As an example, starting with the state vector $|\psi\rangle$ in Eq. C.1,

$$
\begin{aligned}
\hat{R}_z(\delta)|\psi\rangle &= \exp\left(-i\frac{\delta}{2}\sigma_z\right)\begin{pmatrix} \cos(\theta/2) \\ e^{i\varphi}\sin(\theta/2) \end{pmatrix} \\
&= e^{-i\delta/2}\begin{pmatrix} 1 & 0 \\ 0 & e^{i\delta} \end{pmatrix}\begin{pmatrix} \cos(\theta/2) \\ e^{i\varphi}\sin(\theta/2) \end{pmatrix} \\
&= e^{-i\delta/2}\begin{pmatrix} \cos(\theta/2) \\ e^{i(\varphi+\delta)}\sin(\theta/2) \end{pmatrix} \\
&\equiv e^{-i\delta/2}\left[\cos(\theta/2)|1\rangle + e^{i(\varphi+\delta)}\sin(\theta/2)|2\rangle\right],
\end{aligned}
\tag{C.7}
$$

represents a new state (up to an unimportant global phase $-\delta/2$) resulting from a *counterclockwise* rotation of the state vector $|\psi\rangle$ through an angle $\delta$ about the $z$-axis of the Bloch sphere. Equivalently this may be interpreted as a *clockwise* rotation of the Bloch sphere itself through an angle $\delta$. The new Cartesian coordinates can then be derived via the transformation

$$\begin{cases} x' = x\cos(\delta) - y\sin(\delta) \\ y' = x\sin(\delta) + y\cos(\delta) \\ z' = z \end{cases} \quad \text{(C.8)}$$

Together with the identity matrix $I$, the Pauli matrices form a basis for the real vector space of complex Hermitian matrices. Hence the most general Hamiltonian can be written in the form

$$\hat{\mathcal{H}} = \epsilon I + \alpha\sigma_x + \beta\sigma_y + \gamma\sigma_z$$
$$= \epsilon I + E_0 \, \mathbf{n} \cdot \vec{\sigma}, \quad \text{(C.9)}$$

where $\epsilon$, $\alpha$, $\beta$ and $\gamma$ are real coefficients, $E_0 = \sqrt{\alpha^2 + \beta^2 + \gamma^2}$, $\mathbf{n} = (\alpha, \beta, \gamma)/E_0$ is a unitary vector, $\vec{\sigma} = (\sigma_x, \sigma_y, \sigma_z)$ and $E = \epsilon \pm E_0$ represents the system's energy eigenvalues. Consequently, noting that $I$ commutes with all three Pauli matrices, we find that the most general evolution operator can be expressed as

$$\hat{\mathcal{U}} = \exp\left(-\frac{i}{\hbar}\hat{\mathcal{H}}t\right)$$
$$= e^{i\kappa}\exp\left(-\frac{iE_0 t}{\hbar}\mathbf{n}\cdot\vec{\sigma}\right), \quad \text{(C.10)}$$

which, in the Bloch sphere picture, corresponds to a steady rotation of the qubit around the axis $\mathbf{n}$ at a rate $-E_0/\hbar$, up to a global phase $\kappa = -\epsilon t/\hbar$ which may be avoided by setting $\epsilon = 0$.

Although coefficients $\alpha$, $\beta$ and $\gamma$ are system dependant and in most cases cannot be arbitrarily engineered, taking advantage of Euler's rotation theorem, it is often possible to construct a generalized single-qubit operator $\hat{\mathcal{U}}$ by employing a sequence of simpler Bloch vector rotations. More specifically

$$\hat{\mathcal{U}} \equiv \hat{R}_{\mathbf{n}}(\delta) = e^{i\kappa}\hat{R}_i(\delta_3)\hat{R}_j(\delta_2)\hat{R}_i(\delta_1), \quad \text{(C.11)}$$

where $\hat{R}_i$ and $\hat{R}_j$ represent rotations about any two orthogonal axes of the Bloch sphere, as illustrated in Fig. C.2. This decomposition identity has significant implications for the physical realization of unitary quantum gates, since it essentially allows one to implement any arbitrary single-qubit rotation using physical systems whose governing Hamiltonian is naturally confined to a predetermined form with fixed parameter ranges.

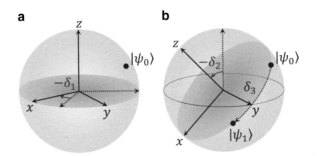

**Fig. C.2** Rotation $\hat{R}_\mathbf{n}(\delta)$ of the state $|\psi_0\rangle$ through an angle $\delta$ about an arbitrary axis $\mathbf{n}$ of the Bloch sphere, accomplished via the sequence $\hat{R}_z(\delta_3)\hat{R}_y(\delta_2)\hat{R}_z(\delta_1)$. The first two rotations $\hat{R}_y(\delta_2)\hat{R}_z(\delta_1)$ rotate the Bloch sphere coordinate system, aligning the initial and final states $|\psi_0\rangle$ and $|\psi_1\rangle$ in the same plane with the $z$-axis oriented along $\mathbf{n}$, before applying the final rotation $\hat{R}_z(\delta_3)$ with $\delta_3 = \delta$

## C.2   Physical Realization

Here we consider the specific case where the qubit basis states $|1\rangle$ and $|2\rangle$ are encoded via a two-level atomic system. The Bloch vector rotations can then be performed by employing the techniques of atom-light interactions developed in earlier sections.

### C.2.1   Rabi Flopping

This scheme is applicable to situations in which it is possible to couple the energy levels corresponding to $|1\rangle$ and $|2\rangle$ directly using an electromagnetic field. The RWA Hamiltonian in Eq. B.37 governing this interaction can be expressed as

$$
\begin{aligned}
\bar{\hat{\mathcal{H}}} &= \frac{\hbar}{2}\left(\mathrm{Re}(\Omega)\sigma_x + \mathrm{Im}(\Omega)\sigma_y + \Delta\sigma_z\right) \\
&= \frac{\hbar}{2}|\Omega_R|\,\mathbf{n}\cdot\vec{\sigma},
\end{aligned}
\tag{C.12}
$$

where $\Omega_R = \sqrt{|\Omega|^2 + \Delta^2}$, $\mathbf{n} = (\mathrm{Re}(\Omega), \mathrm{Im}(\Omega), \Delta)/|\Omega_R|$, $\Delta$ is the detuning from the resonant frequency, $\hbar\Omega = -\mathbf{d}_{21}\cdot\mathbf{E}^{(+)}(z)$ is the interaction energy, $\mathbf{d}_{21}$ is the dipole moment for the $|1\rangle \longrightarrow |2\rangle$ transition (a material dependant constant) and $\mathbf{E}^{(+)}(z)$ is the electric field vector (see Eq. A.9). Given the nature of a short pulse however, the electric field *amplitude* is in fact time varying and hence the interaction energy is more accurately given by the pulse area

$$\Omega = -\frac{1}{\hbar}\,\mathbf{d}_{21}\cdot\int_{-\infty}^{+\infty}\mathbf{E}^{(+)}(z,t)\,dt. \tag{C.13}$$

Now a reference to Eq. C.10 and the preceding discussion reveals that the action of this Hamiltonian for a time $t$ can, in principle, produce general rotations $\hat{R}_{\mathbf{n}}(\Theta = -|\Omega_R|\,t/2)$ of the Bloch vector about any desired axis $\mathbf{n}$ where, $n_x$ and $n_y$ are readily adjusted by appropriately setting the optical phase $\phi$ of the pulse (see Eq. A.14), while $n_z$ depends on the laser detuning.

In practice however, if it is not possible to freely adjust the detuning, generalized Bloch vector rotations can still be achieved using a series of *resonant* laser pulses in accordance with the identity in Eq. C.11. Setting $\Delta = 0$ the RWA Hamiltonian is given by

$$\hat{\mathcal{H}} = \frac{\hbar}{2}\left(\mathrm{Re}(\Omega)\sigma_x + \mathrm{Im}(\Omega)\sigma_y\right)$$

$$= \frac{\hbar}{2}\,|\Omega|\,\mathbf{n}\cdot\vec{\sigma}, \tag{C.14}$$

where $\mathbf{n} = (\cos(\varphi),\sin(\varphi),0)$ and $\tan(\varphi) = \mathrm{Im}(\Omega)/\mathrm{Re}(\Omega)$. This is now equivalent to a rotation $\hat{R}_{\mathbf{n}}(\Theta = -|\Omega|\,t/2)$ of the Bloch vector about an axis $\mathbf{n}$ in the $x$-$y$ plane with azimuthal angle $\varphi$ (which is equal to the pulse phase angle $\phi$ assuming a real valued transition dipole moment). We may now construct a general qubit rotation of the form

$$\hat{R}_{\mathbf{n}}(\delta) = e^{i\kappa}\,\hat{R}_x(\delta_3)\,\hat{R}_y(\delta_2)\,\hat{R}_x(\delta_1), \tag{C.15}$$

performed using three consecutive pulses with phase angles $\varphi = 0,\pi/2,0$ and pulse areas $\Theta = \delta_1,\delta_2,\delta_3$ respectively. Alternatively, when concerned with the application of a series of general rotations, choosing $\hat{R}_z$ instead of $\hat{R}_x$ we find

$$\dots\hat{U}_2\,\hat{U}_1 = \dots\hat{R}_{\mathbf{n}'}(\delta')\,\hat{R}_{\mathbf{n}}(\delta)$$

$$= \dots e^{i\kappa'}\,\hat{R}_z(\delta_3')\,\hat{R}_y(\delta_2')\,\hat{R}_z(\delta_1')\,e^{i\kappa}\,\hat{R}_z(\delta_3)\,\hat{R}_y(\delta_2)\,\hat{R}_z(\delta_1)$$

$$= \dots e^{i(\kappa+\kappa')}\,\hat{R}_z(\delta_3')\,\hat{R}_y(\delta_2')\,\hat{R}_z(\delta_1'+\delta_3)\,\hat{R}_y(\delta_2)\,\hat{R}_z(\delta_1). \tag{C.16}$$

This representation finds a more efficient implementation scheme since the first two rotations $\hat{R}_y(\delta_2)\,\hat{R}_z(\delta_1)$ can be efficiently realized using a single pulse with phase $\varphi = \pi/2 - \delta_1$ and pulse area $\Theta = \delta_2$. Similarly the next two rotations $\hat{R}_y(\delta_2')\,\hat{R}_z(\delta_1'+\delta_3)$ can be performed using another single pulse with phase $\varphi = \pi/2 - \delta_1' - \delta_3$ and pulse area $\Theta = \delta_2'$ and so on.

Particularly important from an application point of view are the so called $\pi$- and $\pi/2$-pulses producing $\Theta = \pi$ and $\Theta = \pi/2$ rotations respectively. The Hadamard gate for example can be produced by a $\pi/2$ rotation about the axis $\mathbf{n} = (1,0,1)/\sqrt{2}$.

## C.2.2   Non-resonance Raman

This scheme is also applicable to situations in which the energy levels corresponding to $|1\rangle$ and $|2\rangle$ are coupled via a non-resonance Raman process involving an intermediate virtual level. Following the adiabatic elimination of the virtual state (see Appendix B.3.2) the effective Hamiltonian in Eq. B.93 can be expressed in the form

$$
\begin{aligned}
\hat{\bar{\mathcal{H}}}_{\text{eff}} &= \frac{\hbar}{2\Delta}\left(|\Omega|^2\, I + \text{Re}(\Omega_T)\sigma_x + \text{Im}(\Omega_T)\sigma_y + \delta\Delta\sigma_z\right) \\
&= \frac{\hbar}{2\Delta}|\Omega|^2\, I + \frac{\hbar}{2\Delta}|\Omega_R|\,\mathbf{n}\cdot\vec{\sigma},
\end{aligned}
\tag{C.17}
$$

where the interaction energies $\Omega_P = \Omega_S = \Omega$ of the two pulses are set to be equal, $\Omega_T = \Omega^2$, $\Delta = \Delta_P + \Delta_S$, $\delta = \Delta_P - \Delta_S$, $\Omega_R = \sqrt{|\Omega_T|^2 + \delta^2\Delta^2}$ and $\mathbf{n} = (\text{Re}(\Omega_T), \text{Im}(\Omega_T), \delta\Delta)/\Omega_R$. It is now easy to see that the Bloch vector rotation resulting from the action of this Hamiltonian can be examined in light of the same analysis pertaining to the Rabi flopping presented in the previous section, noting that the term $(\hbar/2\Delta)/|\Omega|^2\, I$ only contributes as an unimportant global phase.

## C.2.3   STIRAP

The STIRAP scheme described in Appendix B.3.3 is a popular scheme for manipulating qubit systems, due to its proven robustness as an experimental technique for high fidelity population transfer between atomic energy levels. In particular, Kis and Renzon (2002) have shown how arbitrary qubit (Bloch vector) rotations can be implemented using a pair of STIRAP operations.

# Appendix D
# Linear Optical Elements

## D.1 Phase Shifters

*Main Reference (Björk and Söderholm 2003)*

A phase shifter is an optical element with a single input and a single output port, which acts on the input state vector only by introducing a simple phase shift. The phase shifter is characterized by the action of the Hamiltonian in Eq. A.32 on the input modes. For a single mode input state

$$|\psi_{\text{in}}\rangle = \sum_n c_n |n\rangle, \tag{D.1}$$

for example, the Hamiltonian becomes

$$\hat{\mathcal{H}} = \hbar\omega \left( \hat{a}^\dagger \hat{a} + \frac{1}{2} \right), \tag{D.2}$$

and hence

$$|\psi_{\text{out}}\rangle = e^{-i\hat{\mathcal{H}}t/\hbar}|\psi_{\text{in}}\rangle$$
$$= e^{-i\omega t\hat{n}}|\psi_{\text{in}}\rangle$$
$$= \sum_n e^{-i\omega t n} c_n |n\rangle, \tag{D.3}$$

where $\hat{n} = \hat{a}^\dagger \hat{a}$ is the number operator such that $\hat{n}|n\rangle = n|n\rangle$, and we can neglect the unmeasurable constant global phase $\omega/2$. The action of the phase shifter on a coherent input state $|\alpha\rangle$ is given by

$$|\psi_{\text{out}}\rangle = |e^{-i\omega t}\alpha\rangle. \tag{D.4}$$

J. Wang and K. Manouchehri, *Physical Implementation of Quantum Walks*,
Quantum Science and Technology, DOI 10.1007/978-3-642-36014-5,
© Springer-Verlag Berlin Heidelberg 2014

## D.2  Beam Splitters

*Main References (Campos et al. 1989; Campos and Gerry 2005; Yurke et al. 1986; Paris 1996, 1997; Kim et al. 2002; Xiang-bin 2002; Björk and Söderholm 2003)*

A beam splitter is an optical mode mixer which takes an input quantum state $|\psi_{\text{in}}\rangle$ provided by a pair of input ports and produces an output state $|\psi_{\text{out}}\rangle$ exiting via two output ports, where

$$|\psi_{\text{out}}\rangle = \hat{U}_B |\psi_{\text{in}}\rangle, \tag{D.5}$$

and $\hat{U}_B$ represents a unitary transformation. A pure input state can, in its most general form, be written as a special case of Eq. A.38 with only two spatial modes, i.e.

$$|\psi_{\text{in}}\rangle = \sum_i \sum_j c_{ij} |\{n_{\mathcal{M}_i}\}, \{n_{\mathcal{M}'_j}\}\rangle, \tag{D.6}$$

where $n$ represent the photon number in each of the possible frequency and polarization modes $\mathcal{M}_i = \{v, \lambda\}$ incident on port 1 and $\mathcal{M}'_j$ incident on port 2. Although the input state $|\psi_{\text{in}}\rangle$ may in principle be non-separable (e.g. using the output from another beam splitter), it is commonly treated as a simple product state

$$|\psi_{\text{in}}\rangle = |\psi_{A_1}\rangle \otimes |\psi_{A_2}\rangle, \tag{D.7}$$

where $|\psi_{A_1}\rangle$ and $|\psi_{A_2}\rangle$ are the quantum states of the radiation incident on each input port. Taking, for example, single mode Fock states $|n_1\rangle$ and $|n_2\rangle$ entering each port, the input state $|\psi_{\text{in}}\rangle = |n_1, n_2\rangle$.

A beam splitter can be realized as a homogeneous linear and isotropic dielectric medium, in which the polarization vector (dipole moment per unit volume) is aligned with and proportional to the electric field, i.e. $\mathbf{P} = \chi \mathbf{E}$, where $\chi$ is the first order (linear) susceptibility (Paris 1996). Assuming a two-mode frequency matched radiation input (see Eq. A.34)

$$\mathbf{E}(z, t) = \frac{1}{2} \mathbf{E}^{(+)}(z)(\hat{a}_1 + \hat{a}_2) e^{-i\omega t} + \text{c.c.}, \tag{D.8}$$

the Hamiltonian for the beam splitter-radiation coupling is given by

$$\begin{aligned}
\hat{\mathcal{H}} &= \mathbf{P} \cdot \mathbf{E}(z, t) \\
&= \chi \mathbf{E}(z, t) \cdot \mathbf{E}(z, t) \\
&= \frac{1}{4} \chi \left| \mathbf{E}^{(+)}(z) \right|^2 (\underbrace{\hat{a}_1^\dagger \hat{a}_1 + \hat{a}_2^\dagger \hat{a}_2}_{\mathcal{H}_0} + \underbrace{\hat{a}_1 \hat{a}_2^\dagger + \hat{a}_1^\dagger \hat{a}_2}_{\mathcal{H}_{\text{int}}}),
\end{aligned} \tag{D.9}$$

where $\hat{a}_i$ and $\hat{a}_i^\dagger$ represent the annihilation and creation operators for mode $i$ and in the last line we have neglected the high frequency terms (carrying a factor of $\exp(\pm 2i\omega t)$) using the rotating wave approximation (see Appendix B.2.3)). Using the Baker-Campbell-Hausdorff operator identity

$$e^{\xi A} B e^{-\xi A} = B + \xi[A, B] + \frac{\xi^2}{2!}[A, [A, B]] + \dots \qquad (D.10)$$

it is easy to see that the interaction Hamiltonian (see Appendix B.2.4)

$$\hat{\overline{\mathcal{H}}}_I = e^{i\mathcal{H}_0 t/\hbar} \mathcal{H}_{\text{int}} e^{-i\mathcal{H}_0 t/\hbar}$$

$$= \frac{1}{4}\chi \left|\mathbf{E}^{(+)}(z)\right|^2 (\hat{a}_1 \hat{a}_2^\dagger + \hat{a}_1^\dagger \hat{a}_2) \qquad (D.11)$$

contains only those terms involved in the inter-mode photon exchange where the annihilation of one photon in one of the modes leads to the creation of a photon in the other mode and vice versa. The evolution operator in the interaction picture is then given by

$$\hat{U}_B = \exp\left(-\frac{i}{\hbar}\hat{\overline{\mathcal{H}}}_I t\right)$$

$$= \exp\left[\theta\left(\hat{a}_1 \hat{a}_2^\dagger e^{-i\phi} - \hat{a}_1^\dagger \hat{a}_2 e^{i\phi}\right)\right], \qquad (D.12)$$

where

$$\theta e^{-i\phi} = -\frac{i}{4\hbar}\chi \left|\mathbf{E}^{(+)}(z)\right|^2 t, \qquad (D.13)$$

completely characterizes the beam-splitter. It is noteworthy that this evolution operator is in agreement with the often cited work of Campos et al. (1989) in which he derived the beam splitter operator $\hat{B} = \hat{U}_B^\dagger$ (with undetectable phase $\phi_\tau = 0$) on the basis of rotations in the $(2l + 1)$-dimensional Hilbert space of the radiation, generated by the angular momentum operators

$$\hat{L}_1 = \frac{1}{2}(\hat{a}_1^\dagger \hat{a}_2 + \hat{a}_2^\dagger \hat{a}_1), \qquad (D.14)$$

$$\hat{L}_2 = \frac{1}{2i}(\hat{a}_1^\dagger \hat{a}_2 - \hat{a}_2^\dagger \hat{a}_1), \qquad (D.15)$$

$$\hat{L}_3 = \frac{1}{2}(\hat{a}_1^\dagger \hat{a}_1 - \hat{a}_2^\dagger \hat{a}_2). \qquad (D.16)$$

For $l = 1$, for example, the 3D Hilbert space is spanned by the states vectors $|2, 0\rangle$, $|1, 1\rangle$ and $|0, 2\rangle$ and the beam splitter rotation operator $\hat{B}$, finds a $3 \times 3$ matrix representation.

For simple input states however it is actually easiest to initially work in the Heisenberg picture where instead of evolving the input state vector due to $\hat{U}_B$, input operators $\hat{a}_1$ and $\hat{a}_2$ undergo a unitary transformation to produce new operators $\hat{b}_1$ and $\hat{b}_2$ at the beam splitter output. Formally the operators in the Heisenberg picture can be derived via the transformation (Townsend 2000b)

$$\hat{b}_i = \hat{U}_B^\dagger \hat{a}_i \hat{U}_B, \tag{D.17}$$

which leads to the central beam-splitter result

$$\begin{pmatrix} \hat{b}_1(t) \\ \hat{b}_2(t) \end{pmatrix} = M_B \begin{pmatrix} \hat{a}_1(t) \\ \hat{a}_2(t) \end{pmatrix}, \tag{D.18}$$

where

$$M_B = \begin{pmatrix} \cos(\theta) & \sin(\theta)e^{i\phi} \\ -\sin(\theta)^{-i\phi} & \cos(\theta) \end{pmatrix}, \tag{D.19}$$

and we define $t = \cos(\theta)$ and $r = \sin(\theta)$ as the *amplitude* transmission and reflection coefficients, $\tau = \cos^2(\theta)$ and $\rho = \sin^2(\theta)$ are known as the beam splitter *transmittance* and *reflectance* coefficients, and $\phi$ represents a measurable phase difference between the reflected and transmitted components. Here we should also note that any unitary transformation of the type $U(2)$ can be implemented using a lossless beam splitter and a phase shifter at one of the output ports (Cerf et al. 1997).

## D.3   Classical Versus Non-classical Input

*Main References (Björk and Söderholm 2003; Kim et al. 2002; Xiang-bin 2002)*

Let us consider a simple Fock state $|n_1, n_2\rangle$ as the input of a beam splitter. The standard technique for evaluating the output states is to describe the input as the action of creation operators $\hat{a}_1^\dagger$ and $\hat{a}_2^\dagger$ on the vacuum state $|0, 0\rangle$, evolve the creation operators $\hat{a}_1^\dagger \longrightarrow \hat{b}_1^\dagger$ and $\hat{a}_2^\dagger \longrightarrow \hat{b}_2^\dagger$ according to Eq. D.18, and then reconstruct the state using the operators $\hat{b}_1^\dagger$ and $\hat{b}_2^\dagger$. Taking, for example, a single-photon input state $|\psi_{in}\rangle = |1, 0\rangle = \hat{a}_1^\dagger|0, 0\rangle$, together with a 50:50 beam splitter (i.e. $\theta = \pi/4$ and $\phi = \pi/2$), we have

$$\hat{b}_1^\dagger = \frac{1}{\sqrt{2}}(\hat{a}_1^\dagger - i\hat{a}_2^\dagger), \qquad (D.20)$$

$$\hat{b}_2^\dagger = \frac{1}{\sqrt{2}}(\hat{a}_2^\dagger - i\hat{a}_1^\dagger), \qquad (D.21)$$

and therefore

$$
\begin{aligned}
|\psi_{\text{out}}\rangle &= \hat{b}_1^\dagger |0,0\rangle \\
&= \frac{1}{\sqrt{2}}(\hat{a}_1^\dagger - i\hat{a}_2^\dagger)|0,0\rangle \\
&= \frac{1}{\sqrt{2}}|1,0\rangle - \frac{i}{\sqrt{2}}|0,1\rangle.
\end{aligned}
\qquad (D.22)
$$

This is a highly entangled and non-separable state of light due to a non-classical input state. One can then conjecture that in general the entangled output state from a beam splitter requires nonclassicality in the input state (Kim et al. 2002). Xiang-bin (2002) provided a proof for this conjecture.

In contrast to Fock states, coherent states provide a good description of a classical laser beam (see Appendix A.4). For a pair of coherent input states however, we may use a similar approach to evaluate the output state. The coherent state $|\alpha_1, \alpha_2\rangle$ can be defined by the action of the two-mode displacement operator (Glauber 1963; Xiang-bin 2002)

$$\hat{D}(\alpha_1, \alpha_2) = \exp(\alpha_1 \hat{a}_1^\dagger - \alpha_1^* \hat{a}_1 + \alpha_2 \hat{a}_2^\dagger - \alpha_2^* \hat{a}_2) \qquad (D.23)$$

on the vacuum state $|0,0\rangle$. In view of Eq. D.18 it is then easy to show that displacement operators $\hat{D}(\alpha_1, \alpha_2)$ and $\hat{D}(\beta_1, \beta_2)$ at the input and output of the beam-splitter are related according to

$$\begin{pmatrix} \beta_1 \\ \beta_2 \end{pmatrix} = M_B \begin{pmatrix} \alpha_1 \\ \alpha_2 \end{pmatrix}, \qquad (D.24)$$

Noting that $|\alpha|^2 = \langle n \rangle$ is the average photon number in a coherent state, we find that the above relationship is indeed the expected description of a classical beam-splitter with matrix elements $\tau$ and $\rho$ functioning as the standard transmission and reflection coefficients in classical optics. Furthermore, the action of the beam splitter on coherent input states does not lead to any entanglement and the output state is always separable (Kim et al. 2002).

## D.4   Wave Plates

A wave plate or a phase retarder introduces a phase shift between the vertical and horizontal components of the field and thus changes the polarization state of the light wave travelling through it. Wave plates are usually made out of birefringent uniaxial crystals (such as quartz or mica) which are optically anisotropic materials having a refractive index that depends on the polarization and propagation direction of light. Materials with uniaxial anisotropy possess a special axis of symmetry, known as the optical or extraordinary axis. The refractive index $n_e$ along the extraordinary axis is different from $n_o$ experienced along any other ordinary axis perpendicular to it. For negative uniaxial crystals (e.g. calcite and ruby) where $n_e < n_o$, the extraordinary axis is also the fast axis whereas for positive uniaxial crystals (e.g., quartz, magnesium fluoride and rutile) where $n_e > n_o$, the extraordinary axis is the slow axis.

Wave plates are characterized by the amount of relative phase imparted on the two components of input light. Taking advantage of the Jones vector representation in Eq. A.14 the relationship between the input and output polarization states of light can be described as a linear transformation given by

$$\mathbf{J}_{\text{out}} = A_\phi \, \mathbf{J}_{\text{in}}, \tag{D.25}$$

where $A_\phi$ is a 2-by-2 matrix known as the Jones matrix. Consider the case of a linearly polarized light with its $x$ polarization parallel to the optical axis, it can be readily observed that a Jones matrix of the form

$$A_\phi = \begin{pmatrix} e^{i\phi_x} & 0 \\ 0 & e^{i\phi_y} \end{pmatrix}, \tag{D.26}$$

would correctly represent the action of a wave plate by introducing a relative phase shift $\phi = \phi_y - \phi_x$ between the vertical and horizontal components of the field. The amount of this relative phase shift is related to the physical properties of the plate via

$$\phi = \frac{2\pi \Delta n L}{\lambda_0}, \tag{D.27}$$

where $\Delta n = n_y - n_x$ is the material birefringence, $L$ is the wave plate thickness and $\lambda_0$ is the vacuum wavelength of the incident light. In this representation $\phi > 0$ implies that the fast axis lies horizontally alongside the $x$ polarization axis. Conversely $\phi < 0$ means that the wave plate's fast axis is vertical. Wave plates designed to exhibit $|\phi| = \pi/2$ and $\pi$ are commonly referred to as quarter and half-wave plates respectively.

Introducing an angle $\theta$ between the optical and the $x$ polarization axes as in Fig. D.1, we may obtain a new transformation matrix

$$A_\phi(\theta) = R(\theta) \cdot A_\phi \cdot R^{-1}(\theta), \tag{D.28}$$

**Fig. D.1** A wave plate
rotated anti-clockwise by an
angle $\theta$ with respect to the
polarization components $J_x$
and $J_y$ of the input light. $J_x'$
and $J_y'$ are the polarization
components of light in the
wave plate's reference frame

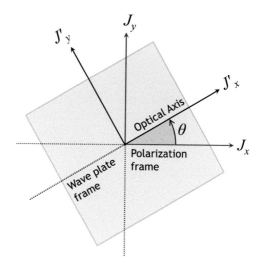

in the polarization reference frame, where $A$ is the original Jones matrix in the wave
plate reference frame, and

$$R(\theta) = \begin{pmatrix} \cos\theta & -\sin\theta \\ \sin\theta & \cos\theta \end{pmatrix}, \tag{D.29}$$

is the usual rotation matrix. It should be noted here that the characteristic behaviour
of a wave plate is fully encapsulated by the relative phase shift $\phi$ it imparts on the
incident beam's polarization components. The absolute phase values $\phi_x$ and $\phi_y$ on
the other hand only contribute to an unimportant global phase which appears in
$A_\phi(\theta)$ as a constant prefactor, meaning they can be adjusted for convenience.

An important special case is the half-wave plate whose associated Jones matrix

$$A_\pi(\theta) = \begin{pmatrix} \cos 2\theta & \sin 2\theta \\ \sin 2\theta & -\cos 2\theta \end{pmatrix}, \tag{D.30}$$

effectively rotates the incoming light's polarization vector by $2\theta$. In particular,
setting $\theta = 0$ provides a convenient method for switching the handedness of
circularly polarized light, while $\theta = 22.5°$ leads to a Hadamard transformation
of the input polarization state. Another configuration with a useful application is a
quarter-wave plate with its optical axis at $\theta = 45°$ with respect to the horizontal. In
this case depending on the orientation of the fast axis (parallel or perpendicular to
the optical axis) $\phi = \pm\pi/2$, resulting in

$$A_{\pi/2}\left(\frac{\pi}{4}\right) = \frac{1}{\sqrt{2}} \begin{pmatrix} 1 & \mp i \\ \mp i & 1 \end{pmatrix}. \tag{D.31}$$

This, together with a horizontal linear polarizer represented by the Jones matrix

$$A_P = \begin{pmatrix} 1 & 0 \\ 0 & 0 \end{pmatrix}, \tag{D.32}$$

are commonly used to create circularly polarized light. It is also possible to generate circular polarization from a linearly polarized input using quarter and half-wave plates at $\theta = 0$ and $\theta = 22.5°$ tilts respectively.

# Appendix E
# Optical Lattices

*Main References (Jaksch 2004; Morsch and Oberthaler 2006; Yin 2006)*

Optical lattices are periodic conservative trapping potentials created by the interference of counter-propagating pairs of laser beams yielding standing laser waves (see Fig. E.1). Neutral atoms (most commonly $^{87}$Rb) are cooled and congregated in the potential minima. The motion of atoms in an optical lattice is closely analogous to that of an electron in a solid state crystal. This situation offers unique opportunities to explore phenomena that were previously accessible only in condensed matter. In contrast to a solid, where the distance between atoms is a few angstroms, the inter-particle spacing in an optical lattice can be many micrometers (Jaksch 2004).

## E.1   Trapping Neutral Atoms

Neutral atom traps (Yin 2006) are often produced using a standard protocol in which the atoms are first collected in a magneto-optical trap (MOT) and cooled down to temperatures in the micro-Kelvin range. The cold atom cloud is then transferred into a conservative trap (which does not reduce the temperature or velocity of the atoms it confines in the way that a MOT does), which may be either magnetic or optical (Morsch and Oberthaler 2006).

An all optical *dipole trap* is formed as a result of the interaction between the atom(s) and the radiation field of a single laser beam, prompting an a.c. Stark shift in the energy levels of the atom(s) (see Appendix B.2.6). The dependance of this a.c. Stark shift on the beam intensity, leads directly to a gradient in the atomic ground state energy, particularly in the Rayleigh range, which is then seen by the atom(s) as the containment potential well.

It is also possible to introduce a second laser field where the resulting interference produces a *dipole lattice trap* or *optical lattice*. Figure E.2 illustrates the distinction between a dipole and a lattice trap. Most importantly, whereas the dipole trap has a

J. Wang and K. Manouchehri, *Physical Implementation of Quantum Walks*,
Quantum Science and Technology, DOI 10.1007/978-3-642-36014-5,
© Springer-Verlag Berlin Heidelberg 2014

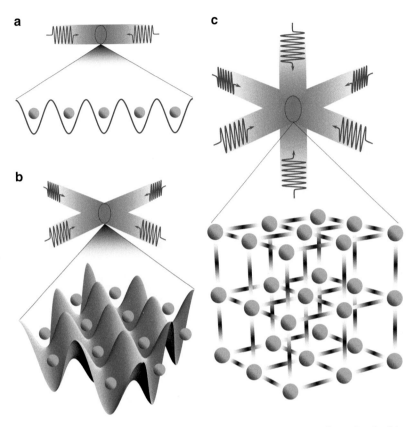

**Fig. E.1** Sets of counter-propagating laser beams producing (**a**) one-dimensional, (**b**) two-dimensional and (**c**) three-dimensional optical lattice configurations for trapping neutral atoms

relatively large (harmonic) confinement, the atoms in an optical lattice are confined to periodic sites that are only half a wavelength apart.

By lowering the harmonic trap depth before switching on the optical lattice however, the atomic cloud undergoes forced evaporative cooling, which can ultimately yield a Bose-Einstein condensate (BEC); a mesoscopic coherent state of matter in which all the atoms show *identical quantum properties*. A BEC is less fragile than one might expect, can be imaged without being destroyed and its interactions with laser light are much stronger than that of a single particle (Jaksch 2004). Such favorable properties have promoted the use of trapped BEC's in numerous experiments. The first experiment demonstrating the loading of a BEC into a 1D optical lattice was carried out by Anderson and Kasevich (1998). In this experiment a magnetically trapped BEC was loaded into a weak vertical optical lattice while at the same time turning off the magnetic trapping almost completely. Later, a 3D lattice geometry was experimentally realized in 2001 by Greiner et al. (2002).

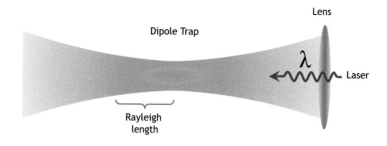

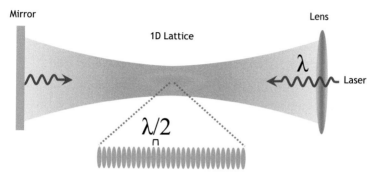

**Fig. E.2** Dipole and lattice traps

Due to quantum tunneling, atoms can move in the optical lattice even if the well depth of the lattice is higher than the atoms' kinetic energy. The motion of atoms in this *superfluid* state is similar to that of electrons in a conductor. In the case of a BEC each atom is spread out over the entire lattice, with long-range phase coherence, depicted in Fig. E.3a. The ensemble will however undergo a quantum phase transition if the interaction energy between the atoms becomes larger than the hopping energy when the well depth is very large (Greiner et al. 2002). In the *Mott insulator* phase, exact numbers of atoms are localized at individual lattice sites and cannot move freely, in a manner similar to the electrons in an insulator (Fig. E.3b). A BEC in the Mott insulator phase exhibits no phase coherence across the lattice. The number of atoms per site, also referred to as the *packing* or *filling factor*, can be experimentally determined to be 1 or more.

**Fig. E.3** Cold atoms in an optical lattice in the (**a**) superfluid and (**b**) Mott insulator phase

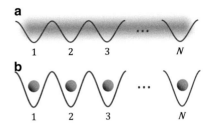

## E.2  Operational Principles

### E.2.1  Electronic Structure of the $^{87}$Rb Atom

The atoms used in the optical trap are often an isotope of rubidium $^{87}$Rb with a nuclear spin $I = \dfrac{3}{2}$. Alkali atoms such as rubidium are used because of their comparatively simple atomic structure where there is only a single outermost electron. In the case of rubidium, this is a $5S$ electron. The energy structure of rubidium's $5S$ and $5P$ shells as illustrated in Fig. E.4. Each shell consists of many degenerate sub-levels (i.e. levels with the same energy) labeled by their angular momentum quantum numbers.

Three fundamental angular momentum vectors, electron spin angular momentum **S**, atomic orbital angular momentum **L** and nuclear spin angular momentum **I** can be added to produce total angular momentum vectors $\mathbf{J} = \mathbf{S} + \mathbf{L}$ and $\mathbf{F} = \mathbf{J} + \mathbf{I}$. These angular momentum vectors and their corresponding quantum numbers $s$, $\ell$, $I$, $j$ and $F$ are associated via the usual relation

$$J \equiv \|\mathbf{J}\| = \sqrt{j(j+1)}\hbar, \tag{E.1}$$

and its projection along an arbitrary $z$-axis

$$J_z = m_j \hbar, \tag{E.2}$$

having $2j + 1$ values $-j, -j+1, \ldots j$, with $j$ ranging from $|s - \ell|$ to $s + \ell$ in integer steps. Furthermore a total angular momentum state can be expanded in terms of the uncoupled basis

$$|jm_j\rangle = \sum_\ell \sum_s \langle \ell m_\ell s m_s | j m_j \rangle |\ell m_\ell\rangle |s m_s\rangle, \tag{E.3}$$

where $\langle \ell m_\ell s m_s | j m_j \rangle \equiv C^{j m_j}_{\ell m_\ell s m_s}$, known as the Clebsch-Gordan coefficients, are nonzero when $m_j = m_\ell + m_s$.

**Fig. E.4** The electronic structure of the 5S and 5P shells in the $^{87}$Rb atom

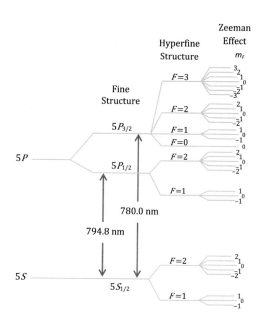

Electrons in the $S$ shells have no orbital angular momentum (i.e. $\ell = 0$) and hence there is no spin-orbit coupling. The total angular momentum quantum number $j$ is then simply equal to the spin angular momentum quantum number $s = \frac{1}{2}$ and the ground state of rubidium is more accurately labeled by $5S_{1/2}$. For the next energy level $5P$ however $\ell = 1$. Hence in the presence of spin-obit coupling, the total angular momentum quantum number $j = \ell \pm \frac{1}{2}$ which leads to a fine structure splitting in the $5P$ shell by lifting the degeneracy between the $5P_{1/2}$ and $5P_{3/2}$ levels corresponding to $j = \frac{1}{2}$ and $j = \frac{3}{2}$. These energy levels are further split into hyperfine sub-levels due to the spin-spin coupling between the electron and the atomic nucleus, where for $5P_{1/2}$ we have $F = 1, 2$ and for $5P_{3/2}$ we have $F = 0, 1, 2, 3$. Each hyperfine level can be further subdivided into $2F + 1$ degenerate sub-levels denoted by quantum numbers $m_F = -F \dots 0 \dots F$. This degeneracy can be lifted by the application of an external magnetic field (Zeeman effect) and the $z$-axis is fixed in the direction of the field.

## E.2.2  Activating the Trapping Potential

*Main References (Deutsch and Jessen 1998; Jessen and Deutsch 1996; Dalibard and Cohen-Tannoudji 1989; Metcalf and der Straten 1999b; Oberthaler 2007)*

A 1D lattice is produced by a pair of counter-propagating plane laser waves which, as depicted in Fig. E.5a, travel in directions $\pm \hat{\mathbf{z}}$ with an angle $\theta$ between their linear polarizations, a configuration that is referred to as the 1D "lin-angle-lin" or "lin $\angle$ lin", and is reduced to the familiar "lin $\perp$ lin" lattice when $\theta = \pi/2$. Assuming electric fields with equal amplitudes $E_0$, the total field is then given by (see Eq. A.6)

$$\mathbf{E}_{\text{total}}(z,t) = \frac{1}{2}\mathbf{E}_{\text{total}}^{(+)}(z)e^{-i\omega t} + \text{ c.c.,} \qquad (E.4)$$

where

$$
\begin{aligned}
\mathbf{E}_{\text{total}}^{(+)}(z) &= E_0 e^{i\phi/2}\begin{pmatrix}\cos(\frac{\theta}{2})\\ \sin(\frac{\theta}{2})\end{pmatrix}e^{ikz} + E_0 e^{-i\phi/2}\begin{pmatrix}\cos(-\frac{\theta}{2})\\ \sin(-\frac{\theta}{2})\end{pmatrix}e^{-ikz}\\
&= \sqrt{2}E_0[\cos(kz+\phi/2+\theta/2)\,\mathbf{e}_+ + \cos(kz+\phi/2-\theta/2)\,\mathbf{e}_-],\\
&= \mathbf{E}_-^{(+)}(z) + \mathbf{E}_+^{(+)}(z), \qquad (E.5)
\end{aligned}
$$

$k = \omega/2\pi c$ is the laser wave vector, $\phi$ is the relative phase between the laser pair and the field resulting from the interaction is expressed as the superposition of left circularly polarized $\sigma^-$ and right circularly polarized $\sigma^+$ standing waves (see Eq. A.17).

Both trapping lasers are coupled to the D2 transition of the rubidium atoms i.e. $5S_{1/2} \longrightarrow 5P_{3/2}$ (with $\lambda \approx 780\,\text{nm}$) which can be described as a $j = \frac{1}{2} \longrightarrow j' = \frac{3}{2}$ transition. As depicted in Fig. E.6 the presence of ground and excited spin states labeled by $|g_{m_j}\rangle$ and $|e_{m_{j'}}\rangle$, gives rise to six possible transitions involving the absorption of $-\hbar$, $0$ or $+\hbar$ units of angular momentum. These transitions (or a combination thereof) may therefore be invoked by interacting the atom with a field that is respectively left circularly polarized, linearly polarized or right circularly polarized.

To determine the interaction energy $\Omega$ associated with a transition involving $|g_{m_j}\rangle$ and $|e_{m_{j'}}\rangle$ the electric dipole moment between these states must be calculated (see Eq. B.27). This is greatly simplified by the Wigner-Eckart theorem stating

$$\langle jm_j|\hat{T}_q^k|j'm_{j'}\rangle = \langle j|\hat{T}^k|j'\rangle C_{jm_jkq}^{j'm_{j'}}, \qquad (E.6)$$

where $\hat{T}_q^k$ is a rank $k$ *spherical* tensor operator with $q$ being an integer in the range $-k, -k+1\ldots, k-1, k$, $C_{jm_jkq}^{j'm_{j'}}$ is a Clebsch-Gordan coefficient (see Eq. E.3) and $\langle j|\hat{T}^k|j'\rangle$ is independent of the projection quantum numbers $m_{j'}, m_j$ and $q$. Hence, to compute the dipole moment vector in the presence of spin-obit coupling, Eq. B.18 is re-written in the form

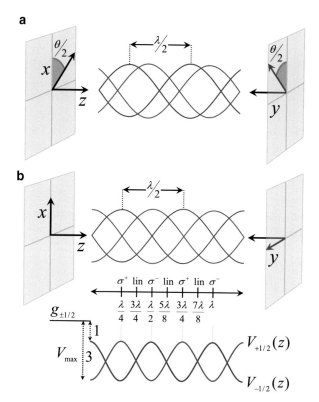

**Fig. E.5** A pair of counter-propagating lasers in the (**a**) lin $\angle$ lin and (**b**) lin $\perp$ lin configurations producing a superposition of standing *left* circularly polarized $\sigma^-$ (*red*) and *right* circularly polarized $\sigma^+$ (*blue*) fields. In the lin $\perp$ lin configuration the superposition leads to spatially varying polarizations in the light field, changing periodically from $\sigma^+$, to linearly polarized, $\sigma^-$, corresponding to a periodic trapping potential $V(x)$

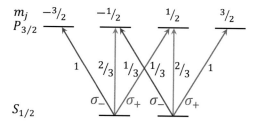

**Fig. E.6** Fine structure of the rubidium atom and various $j = 1/2 \longrightarrow j' = 3/2$ transitions due to $\sigma^-$-, linearly- and $\sigma^+$-polarized field along with their relative strengths given by the Clebsch-Gordan coefficients

$$\mathbf{d}_{jj'} \equiv \langle j |\hat{\mathbf{d}}| j'\rangle = \langle jm_j |\hat{\mathbf{d}}| j'm_{j'}\rangle = -e\langle j |\hat{\mathbf{r}}| j'\rangle \, C^{j'm_{j'}}_{jm_j qk}, \qquad (E.7)$$

where the dipole moment vector operator $\hat{\mathbf{d}}$ is a spherical tensor operator of rank $k = 1$ with $q = -1, 0, 1$, quantum numbers $j = \dfrac{1}{2}$ and $j' = \dfrac{3}{2}$ correspond to the ground and excited states $5S_{1/2}$ and $5P_{3/2}$ respectively, $C^{j'm_{j'}}_{jm_j qk}$ is the Clebsch-Gordan coefficient of the transition, and the term $-e\langle j |\hat{\mathbf{r}}| j'\rangle$ accounts for the radial overlap between the ground and excited state wave-functions which is independent of the projection quantum numbers $m_j$, $m_{j'}$ and $q$, and is hence the same for all transitions.

Consequently by simply comparing the Clebsch-Gordan coefficients, hereafter denoted by $C_{m_j m_{j'}}$ for shorthand, one can immediately determine the relative strength of each transition due to individual field components $\mathbf{E}^{(+)}_{\pm}(z)$ and its associated light shift (see Eq. B.74)

$$\Delta \mathcal{E}_{\pm}(m_j, m_{j'}) = \frac{\Delta \left| \mathbf{d}_r \cdot \mathbf{E}^{(+)}_{\pm}(z) \right|^2 C^2_{m_j m_{j'}}}{4\hbar \left( \Delta^2 + \Gamma^2/4 \right)}. \qquad (E.8)$$

The overall potential $V_{\pm 1/2}$ experienced by an atom in the ground state $|g_{\pm 1/2}\rangle$ is the sum of all light shifts due to every transition from that ground state, when interacting with the total field $\mathbf{E}^{(+)}_{\text{total}}(z)$. In other words

$$V_{1/2} = \Delta \mathcal{E}_{+}\left(\tfrac{1}{2}, \tfrac{3}{2}\right) + \Delta \mathcal{E}_{-}\left(\tfrac{1}{2}, -\tfrac{1}{2}\right), \quad \text{and} \qquad (E.9)$$

$$V_{-1/2} = \Delta \mathcal{E}_{-}\left(-\tfrac{1}{2}, -\tfrac{3}{2}\right) + \Delta \mathcal{E}_{+}\left(-\tfrac{1}{2}, \tfrac{1}{2}\right), \qquad (E.10)$$

which gives

$$V_{\pm 1/2}(z, \theta) = V_{\pm}(z, \theta) + \frac{1}{3} V_{\mp}(z, \theta), \qquad (E.11)$$

where

$$V_{\pm}(z, \theta) = V_{\max} \cos^2(kz + \phi/2 \pm \theta/2), \qquad (E.12)$$

and the unit light shift

$$V_{\max} = \frac{\Delta \, E_0^2 \, \|\mathbf{d}_r\|^2}{2\hbar \left( \Delta^2 + \Gamma^2/4 \right)}. \qquad (E.13)$$

Expressing the lattice potential in the form

$$V_{\pm 1/2}(z, \theta) = \frac{1}{2} V_{\max} \cos(2kz + \phi \pm \theta) + \frac{1}{6} V_{\max} \cos(2kz + \phi \mp \theta) + \frac{2}{3} V_{\max}, \quad (E.14)$$

facilitates the specialization of $V_{\pm 1/2}(z, \theta)$ for various polarization angles. In particular, with $\phi = 0$, two useful trapping potentials emerge, one for $\theta = 0$ where

$$V_{\pm 1/2}(z, 0) = \frac{4}{3} V_{\text{max}} \cos^2(kz), \tag{E.15}$$

and the other for the lin $\perp$ lin configuration where

$$V_{\pm 1/2}(z, \pi/2) = \frac{2}{3} V_{\text{max}} \cos^2(kz \pm \pi/4) + \frac{1}{3} V_{\text{max}}. \tag{E.16}$$

As illustrated in Fig. E.5b in the latter case the maximum light shift occurs at the point where $\mathbf{E}_{\text{total}}^{(+)}(z)$ is purely right (left) circularly polarized, corresponding to the transition $m_j = \pm\frac{1}{2} \longrightarrow m_{j'} = \pm\frac{3}{2}$ of the atom in the ground state $|g_{\pm 1/2}\rangle$. The amount of light shift is then halved where the polarization becomes linear, corresponding to the transition $m_j = \pm\frac{1}{2} \longrightarrow m_{j'} = \pm\frac{1}{2}$, and finally reduced to its minimum value $V_{\text{max}}/3$ where the field is entirely left (right) circularly polarized, corresponding to the transition $m_j = \pm\frac{1}{2} \longrightarrow m_{j'} = \mp\frac{1}{2}$. Depending on their internal state, the atoms will naturally congregate at the minima of either periodic potentials.

It is instructive here to also note the impact of the laser frequency on the quality of the trapping potential. It follows from Eqs. B.73 and B.74 that for $|\Delta| \gg \Gamma$, losses due to spontaneous emission out of $5P_{3/2}$ scale as $|E_0|^2 \Gamma/\Delta^2$, whereas the potential depth scales as $|E_0|^2/\Delta$. Therefore a far off-resonant intense laser beam is able to produce a deep enough potential with sufficiently small atomic loss to effectively confine the atoms.

## E.3  Spin Dependant Transport

*Main References (Nakahara and Ohmi 2008; Mandel et al. 2003a)*

As evident from Eq. E.11, when the counter propagating beams have different polarizations, i.e. $\theta > 0$, the resulting potential is dependent on the spin state of the atom. As depicted in Fig. E.7, the objective of spin dependant transport is to move the two potentials (and the atoms confined within them) with respect to one another by changing the polarization angle $\theta$. In order to do so two internal spin states of the atom should be used, where one spin state dominantly experiences the $V_-(z, \theta)$ potential and the other spin state mainly experiences the $V_+(z, \theta)$ dipole force potential (Mandel et al. 2003a). Such a situation can be realized in rubidium by tuning the wavelength of the optical lattice laser to a value of $\lambda = 785$ nm between the fine structure splitting of the rubidium D1 and D2 transitions. This is illustrated in Fig. E.8a where transitions correspond to atomic excitations due to the $\sigma^-$-polarized component of the field denoted by $\mathbf{E}_-^{(+)}(z)$. Figure E.8b depicts the corresponding potentials

**Fig. E.7** Spin dependant transport of neutral atoms in an optical lattice by changing the relative polarization angle $\theta$ between the counter-propagating lasers. Atoms in their internal ground and excited states see two different potentials which move relative to each other in response to a change in $\theta$

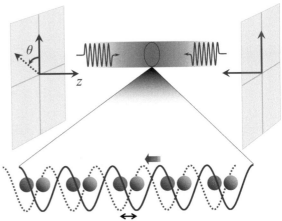

**Fig. E.8** (a) Fine structure of the rubidium atom and excitation of the atom with a $\sigma^-$-polarized field. The ground state with $m_j = 1/2$ may be excited to states $P_{1/2}$ and $P_{3/2}$ both with $m_{j'} = -1/2$, corresponding to D1 and D2 transitions respectively. The ground state $m_j = -1/2$ may be excited to $P_{3/2}$ only. (b) Light shift of the ground states $m_j = 1/2$ (*solid*) and $m_j = -1/2$ (*dashed*). The light shift for the $m_j = 1/2$ vanishes at $\omega = \omega_0$ (Adapted from Nakahara and Ohmi (2008))

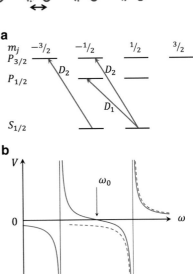

$$V_{1/2} = \Delta\mathcal{E}_- (D2, 1/2, -1/2) + \Delta\mathcal{E}_- (D1, 1/2, -1/2), \quad \text{and}$$

$$V_{-1/2} = \Delta\mathcal{E}_- (D2, -1/2, -3/2), \tag{E.17}$$

using Eq. E.8 and assuming $\Gamma \approx 0$, in which case

$$\Delta\mathcal{E}_+(D1) \propto \frac{1}{\omega - \omega_1} \quad \text{and} \quad \Delta\mathcal{E}_+(D2) \propto \frac{1}{\omega - \omega_2}, \tag{E.18}$$

where $\omega_1$ and $\omega_2$ denote the resonant frequencies associated with D1 and D2 transitions respectively. It can be seen in Fig. E.8b that the potential $V_{1/2}$ vanishes when $\omega = \omega_0$ while $V_{-1/2}$ remains appreciable. A similar argument involving the

$\mathbf{E}_+^{(+)}(z)$ component shows that $V_{-1/2}$ vanishes also when $\omega = \omega_0$ while $V_{1/2}$ remains appreciable. Consequently we find that under the influence of the total field $\mathbf{E}_{\mathrm{total}}^{(+)}(z)$ tuned to the magic frequency $\omega_0$

$$V_{\pm 1/2}(z, \theta) = V_\pm(z, \theta)$$

$$= V_{\max} \cos^2(kz \pm \theta/2), \qquad (E.19)$$

where $V_{\max}$ is given by Eq. E.13 and the phase angle $\phi = 0$. The separation between the two potentials

$$\Delta z = \frac{\theta}{\pi} \frac{\lambda_z}{2} \qquad (E.20)$$

can now be readily controlled by changing the polarization angle $\theta$. When increasing $\theta$, both potentials shift in opposite directions and overlap again when $\theta = n\pi$, with $n$ being an integer.

Considering the two ground states as forming a qubit system, it is typical to choose our basis states to be magnetically sensitive, such as $|0\rangle = |F = 1, m_F = -1\rangle$ and $|1\rangle = |F = 2, m_F = -2\rangle$ employed by Mandel et al. (2003a). Since the total angular momentum quantum number $F = j + I$, these basis states can be expressed in terms of the their uncoupled basis (see Eq. E.3)

$$|Fm_F\rangle = \sum_I \sum_j C_{m_I m_j m_F}^{IjF} |Im_I\rangle |jm_j\rangle, \qquad (E.21)$$

which evaluates to

$$|F = 2, m_F = -2\rangle = \left|\frac{3}{2}, \frac{3}{2}\right\rangle \left|\frac{1}{2}, \frac{1}{2}\right\rangle, \quad \text{and} \qquad (E.22)$$

$$|F = 1, m_F = -1\rangle = -\frac{\sqrt{3}}{2} \left|\frac{3}{2}, -\frac{3}{2}\right\rangle \left|\frac{1}{2}, \frac{1}{2}\right\rangle + \frac{1}{2} \left|\frac{3}{2}, -\frac{1}{2}\right\rangle \left|\frac{1}{2}, -\frac{1}{2}\right\rangle. \qquad (E.23)$$

Therefore the dipole potential experienced by an atom in state $|1\rangle$ or $|0\rangle$ is given by

$$V_1(z, \theta) = V_+(z, \theta), \quad \text{and} \qquad (E.24)$$

$$V_0(z, \theta) = \frac{3}{4} V_+(z, \theta) + \frac{1}{4} V_-(z, \theta). \qquad (E.25)$$

# Bibliography

Abal, G., Donangelo, R., Fort, H.: Annals of the 1st Workshop on Quantum Computation and Information, pp. 189–200 (2006a). arXiv:quant-ph/0709.3279v1

Abal, G., Donangelo, R., Romanelli, A., Siri, R.: Effects of non-local initial conditions in the quantum walk on the line. Physica A **371**, 1 (2006b)

Abal, G., Siri, R., Romanelli, A., Donangelo, R.: Quantum walk on the line: entanglement and nonlocal initial conditions. Phys. Rev. A **73**, 042302 (2006c)

Abala, G., Donangeloa, R., Severoa, F., Siria, R.: Decoherent quantum walks driven by a generic coin operation. Physica A **387**, 335 (2008)

Adamczak, W., Peter, K.A., Tamon, H.C.: (2003). arXiv:quant-ph/0308073

Agarwal, G.S., Pathak, P.K.: Quantum random walk of the field in an externally driven cavity. Phys. Rev. A **72**, 033815 (2005)

Agliari, E., Blumen, A., Mülken, O.: Quantum-walk approach to searching on fractal structures. Phys. Rev. A **82**, 012305 (2010)

Aharonov, Y., Davidovich, L., Zagury, N.: Quantum random walks. Phys. Rev. A **48**, 1687 (1993)

Aharonov, D., Ambainis, A., Kempe, J., Vazirani, U.: STOC'01: Proceedings of the 33rd Annual ACM Symposium on Theory of Computing, pp. 50–59. ACM (2001). arXiv:quant-ph/0012090

Alagić, G., Russell, A.: Decoherence in quantum walks on the hypercube. Phys. Rev. A **72**, 062304 (2005)

Allen, L., Beijersbergen, M.W., Spreeuw, R.J.C., Woerdman, J.P.: Optical angular momentum of light and the transformation of Laguerre-Gaussian laser modes. Phys. Rev. A **45**, 8185 (1992)

Ambainis, A.: Quantum walks and their algorithmic applications. Int. J. Quantum Inf. **1**, 507 (2003)

Ambainis, A.: FOCS04: Proceedings of the 45th Annual IEEE Symposium on Foundations of Computer Science, pp. 22–31. IEEE Computer Society (2004). arXiv:quant-ph/0311001

Ambainis, A., Bach, E., Nayak, A., Vishwanath, A., Watrous, J.: STOC'01: Proceedings of the 33rd Annual ACM Symposium on Theory of Computing, pp. 50–59. ACM (2001)

Ambainis, A., Kempe, J., Rivosh, A.: SODA'05: Proceedings of the 16th Annual ACM-SIAM Symposium on Discrete Algorithms, pp. 1099–1108. Society for Industrial and Applied Mathematics (2005). arXiv:quant-ph/0402107

Anderson, P.W.: Absence of diffusion in certain random lattices. Phys. Rev. **109**, 1492 (1958)

Anderson, B.P., Kasevich, M.A.: Macroscopic quantum interference from atomic tunnel arrays. Science **282**, 1686 (1998)

Anteneodo, C., Morgado, W.A.M.: Critical scaling in standard biased random walks. Phys. Rev. Lett. **99**, 180602 (2007)

J. Wang and K. Manouchehri, *Physical Implementation of Quantum Walks*,
Quantum Science and Technology, DOI 10.1007/978-3-642-36014-5,
© Springer-Verlag Berlin Heidelberg 2014

Aslangul, C.: Quantum dynamics of a particle with a spin-dependent velocity. J. Phys. A **38**, 1 (2005)

Babai, L., Kantor, W.M., Luks, E.M.: Proceedings of the 24th Annual Symposium on Foundations of Computer Science SFCS '83, pp. 162–171. IEEE Computer Society (1983)

Bach, E., Coppersmith, S., Goldschen, M.P., Joynt, R., Watrous, J.: One-dimensional quantum walks with absorbing boundaries. J. Comput. Syst. Sci. **69**, 562 (2004)

Bañuls, M.C., Navarrete, C., Pérez, A., Roldán, E., Soriano, J.C.: Quantum walk with a time-dependent coin. Phys. Rev. A **73**, 062304 (2006)

Bar-Yossef, Z., Gurevich, M.: WWW'06: Proceedings of the 15th International Conference on World Wide Web, pp. 367–376. ACM (2006)

Bazhenov, V.Y., Vasnetsov, M.V., Soskin, M.S.: Laser beams with screw dislocations in their wavefront. JETP Lett. **52**, 429 (1990)

Benenti, G., Casati, G., Strini, G.: Principles of Quantum Computation and Information: Basic Concepts, vol. 1. World Scientific Publishing, Singapore (2004). Chapter 3

Bénichou, O., Coppey, M., Moreau, M., Suet, P.-H., Voituriez, R.: Optimal search strategies for hidden targets. Phys. Rev. Lett. **94**, 198101 (2005)

Berg, B.A.: Locating global minima in optimization problems by a random-cost approach. Nature **361**, 708 (1993)

Bergamini, S., Darquié, B., Jones, M., Jacubowiez, L., Browaeys, A., Grangier, P.: Holographic generation of microtrap arrays for single atoms by use of a programmable phase modulator. J. Opt. Soc. Am. B **21**, 1889 (2004)

Bergmann, K., Theuer, H., Shore, B.W.: Coherent population transfer among quantum states of atoms and molecules. Rev. Mod. Phys. **70**, 1003 (1998)

Berry, S.D., Wang, J.B.: Two-particle quantum walks: entanglement and graph isomorphism testing. Phys. Rev. A **83**, 042317 (2011)

Björk, G., Söderholm, J.: The Dirac-Notation in Quantum Optics, Technical Report, Department of Microelectronics and Information Technology, KTH Electrum, Kista (2003)

Blanchard, P., Hongler, M.-O.: Quantum random walks and piecewise deterministic evolutions. Phys. Rev. Lett. **92**, 120601 (2004)

Bouwmeester, D., Marzoli, I., Karman, G.P., Schleich, W., Woerdman, J.P.: Optical Galton board. Phys. Rev. A **61**, 013410 (1999)

Bracken, A.J., Ellinas, D., Smyrnakis, I.: Free-Dirac-particle evolution as a quantum random walk. Phys. Rev. A **75**, 022322 (2007)

Brion, E., Pedersen, L.H., Mølmer, K.: Adiabatic elimination in a lambda system. J. Phys. A **40**, 1033 (2007)

Broome, M.A., Fedrizzi, A., Lanyon, B.P., Kassal, I., Aspuru-Guzik, A., White, A.G.: Discrete single-photon quantum walks with tunable decoherence. Phys. Rev. Lett. **104**, 153602 (2010)

Brun, T.A., Carteret, H.A., Ambainis, A.: Quantum to classical transition for random walks. Phys. Rev. Lett. **91**, 130602 (2003a)

Brun, T.A., Carteret, H.A., Ambainis, A.: Quantum walks driven by many coins. Phys. Rev. A **67**, 052317 (2003b)

Brun, T.A., Carteret, H.A., Ambainis, A.: Quantum random walks with decoherent coins. Phys. Rev. A **67**, 032304 (2003c)

Cai, J.-Y., Fürer, M., Immerman, N.: An optimal lower bound on the number of variables for graph identification. Combinatorica **12**, 389 (1992)

Calarco, T., Dorner, U., Julienne, P.S., Williams, C.J., Zoller, P.: Quantum computations with atoms in optical lattices: marker qubits and molecular interactions. Phys. Rev. A **70**, 012306 (2004)

Campos, R.A., Gerry, C.C.: Permutation-parity exchange at a beam splitter: application to Heisenberg-limited interferometry. Phys. Rev. A **72**, 065803 (2005)

Campos, R.A., Saleh, B.E.A., Teich, M.C.: Quantum-mechanical lossless beam splitter: SU(2) symmetry and photon statistics. Phys. Rev. A **40**, 1371 (1989)

Carneiro, I., Loo, M., Xu, X., Girerd, M., Kendon, V., Knight, P.L.: Entanglement in coined quantum walks on regular graphs. New J. Phys. **7**, 156 (2005)

Ceperley, D., Alder, B.: Quantum Monte Carlo. Science **231**, 555 (1986)

Cerf, N.J., Adami, C., Kwiat, P.G.: Optical simulation of quantum logic. Phys. Rev. A **57**, 1477 (1997)

Cerf, N.J., Adami, C., Kwiat, P.G.: Optical simulation of quantum logic. Phys. Rev. A **57**, 1477 (1998)

Chandrashekar, C.M.: Implementing the one-dimensional quantum (Hadamard) walk using a Bose-Einstein condensate. Phys. Rev. A **74**, 032307 (2006)

Childs, A.M.: Phys. Rev. Lett. **102**, 180501 (2009)

Childs, A.M.: On the relationship between continuous- and discrete-time quantum walk. Commun. Math. Phys. **294**, 581 (2010)

Childs, A., Goldstone, J.: Spatial search and the Dirac equation. Phys. Rev. A **70**, 042312 (2004a)

Childs, A., Goldstone, J.: Spatial search by quantum walk. Phys. Rev. A **70**, 022314 (2004b)

Childs, A., Farhi, E., Gutmann., S.: An example of the difference between quantum and classical random walks. Quantum Inf. Process. **1**, 35 (2002)

Childs, A., Cleve, R., Deotto, E., Farhi, E., Gutmann, S., Spielman, D.: STOC'03: Proceedings of the 35th Annual ACM Symposium on Theory of Computing, pp. 59–68 ACM (2003). arXiv:quant-ph/0209131

Cho, J.: Addressing individual atoms in optical lattices with standing-wave driving fields. Phys. Rev. Lett. **99**, 020502 (2007)

Choi, W., Lee, M., Lee, Y.-R., Park, C., Lee, J.-H., Ana, K., Fang-Yen, C., Dasari, R.R., Feld, M.S.: Calibration of second-order correlation functions for nonstationary sources with a multistart, multistop time-to-digital converter. Rev. Sci. Inst. **76**, 083109 (2005)

Cives-esclop, A., Luis, A., Sánchez-Soto, L.L.: Influence of field dynamics on Rabi oscillations: beyond the standard semiclassical Jaynes-Cummings model. J. Mod. Opt. **46**, 639 (1999)

Clementi, E., Raimondi, D.L., Reinhardt, W.: Atomic screening constants from SCF functions. II. atoms with 37 to 86 electrons. J. Chem. Phys. **47**, 1300 (1967)

Côté, R., Russell, A., Eyler, E.E., Gould, P.L.: Quantum random walk with Rydberg atoms in an optical lattice. New J. Phys. **8**, 156 (2006)

Dalibard, J., Cohen-Tannoudji, C.: Laser cooling below the Doppler limit by polarization gradients: simple theoretical models. J. Opt. Soc. Am. B **100**, 2023 (1989)

de la Torre, A.C., Mártin, H.O., Goyeneche, D.: Quantum diffusion on a cyclic one-dimensional lattice. Phys. Rev. E **68**, 031103 (2003)

Deutsch, I.H., Jessen, P.S.: Quantum-state control in optical lattices. Phys. Rev. A **57**, 1972 (1998)

Di, T., Hillery, M., Zubairy, M.S.: Cavity QED-based quantum walk. Phys. Rev. A **70**, 032304 (2004)

Dinneen, C., Wang, J.B.: Phase dynamics in charge Qubits. J. Comput. Theor. Nanosci. **5**, 1 (2008)

Do, B., Stohler, M.L., Balasubramanian, S., Elliott, D.S., Eash, C., Fischbach, E., Fischbach, M.A., Mills, A., Zwickl, B.: Experimental realization of a quantum quincunx by use of linear optical elements. Opt. Soc. Am. B **22**, 020499 (2005)

Douglas, B.L.: (2011). arXiv:1101.5211 and private communication

Douglas, B.L., Wang, J.B.: A classical approach to the graph isomorphism problem using quantum walks. J. Phys. A **41**, 075303 (2008)

Douglas, B.L., Wang, J.B.: Efficient quantum circuit implementation of quantum walks. Phys. Rev. A **79**, 052335 (2009)

Du, J., Li, H., Xu, X., Shi, M., Wu, J., Zhou, X., Han, R.: Experimental implementation of the quantum random-walk algorithm. Phys. Rev. A **67**, 042316 (2003)

Dumke, R., Volk, M., Müther, T., Buchkremer, F.B.J., Birkl, G., Ertmer, W.: Micro-optical realization of arrays of selectively addressable dipole traps: a scalable configuration for quantum computation with atomic qubits. Phys. Rev. Lett. **89**, 097903 (2002)

Dur, W., Raussendorf, R., Kendon, V.M., Briegel, H.J.: Quantum walks in optical lattices. Phys. Rev. A **66**, 052319 (2002)

Eckert, K., Mompart, J., Birkl, G., Lewenstein, M.: One- and two-dimensional quantum walks in arrays of optical traps. Phys. Rev. A **72**, 012327 (2005)

Einstein, A.: Quantentheorie der Strahlung. Phys. Z. **18**, 121 (1917)

Emms, D., Hancock, E.R., Severini, S., Wilson, R.: A matrix representation of graphs and its spectrum as a graph invariant. Electron. J. Comb. **13**, R34 (2006)

Englert, B.-G., Kurtsiefer, C., Weinfurter, H.: Universal unitary gate for single-photon two-qubit states. Phys. Rev. A **63**, 032303 (2001)

Farhi, E., Gutmann, S.: Quantum computation and decision trees. Phys. Rev. A **58**, 915 (1998)

Feldman, E., Hillery, M.: Scattering theory and discrete-time quantum walk. Phys. Lett. A **324**, 277 (2004)

Feynman, R.P.: Quantum mechanical computers. Opt. News **11**, 11 (1985)

Foden, C.L., Talyanskii, V.I., Milburn, G.J., Leadbeater, M.L., Pepper, M.: High-frequency acousto-electric single-photon source. Phys. Rev. A **62**, 011803 (2000)

Foot, C.J.: Atomic Physics. Oxford University Press, Oxford (2005). Chapter 7

Fox, M.: Quantum Optics: An Introduction. Oxford University Press, Oxford (2007). Chapter 13

Francisco, D., Iemmi, C., Paz, J., Ledesma, S.: Optical simulation of the quantum Hadamard operator. Opt. Commun. **268**, 340 (2006a)

Francisco, D., Iemmi, C., Paz, J.P., Ledesma, S.: Simulating a quantum walk with classical optics. Phys. Rev. A **74**, 052327 (2006b)

Galton, F.: Typical laws of heredity. Nature **15**, 492 (1877)

Gamble, J.K., Friesen, M., Zhou, D., Joynt, R., Coppersmith, S.N.: Two-particle quantum walks applied to the graph isomorphism problem. Phys. Rev. A **81**, 052313 (2010)

Gaubatz, U., Rudecki, P., Schiemann, S., Bergmann, K.: Population transfer between molecular vibrational levels by stimulated Raman scattering with partially overlapping laser fields. A new concept and experimental results. J. Chem. Phys. **92**, 5363 (1990)

Gerhardt, H., Watrous, J.: RANDOM'03: Proceedings of the 7th International Workshop on Randomization and Approximation Techniques in Computer Science, Princeton. Lecture Notes in Computer Science, vol. 2764, pp. 290–301. Springer, New York (2003)

Gericke, T., Würtz, P., Reitz, D., Langen, T., Ott, H.: High-resolution scanning electron microscopy of an ultracold quantum gas. Nat. Phys. **4**, 949 (2008)

Gerry, C.C.: Dynamics of SU(1,1) coherent states. Phys. Rev. A **31**, 2226 (1985)

Gershenfeld, N.A., Chuang, I.L.: Bulk spin-resonance quantum computation. Science **275**, 350 (1997)

Glauber, R.J.: Coherent and incoherent states of the radiation field. Phys. Rev. Lett. **131**, 2767 (1963)

Godsil, C., Royle, G.F.: Algebraic Graph Theory. Graduate Texts in Mathematics, vol. 207. Springer, New York (2001)

Gorshkov, A.V., Jiang, L., Greiner, M., Zoller, P., Lukin, M.D.: Coherent quantum optical control with subwavelength resolution. Phys. Rev. Lett. **100**, 093005 (2008)

Green, R., Wang, J.B.: Quantum dynamics of gate operations on charge qubits. J. Comput. Theor. Nanosci. **1**, 1 (2005)

Greiner, M., Mandel, O., Esslinger, T., Hänsch, T.W., Bloch, I.: Quantum phase transition from a superfluid to a Mott insulator in a gas of ultracold atoms. Nature **415**, 39 (2002)

Greiner, M., Regal, C.A., Jin, D.S.: Emergence of a molecular Bose-Einstein condensate from a Fermi Gas. Nature **426**, 537 (2003)

Griffiths, D.J.: Introduction to Electrodynamics, 3rd edn. Prentice-Hall, Upper Saddle River (1999). Chapter 10

Grover, L.K.: STOC'96: Proceedings of the 28th Annual ACM Symposium on Theory of Computing, Philadephia, pp. 212–219. ACM Press, New York (1996)

Gurvitz, S.A., Fedichkin, L., Mozyrsky, D., Berman, G.P.: Relaxation and the zeno effect in qubit measurements. Phys. Rev. Lett. **91**, 066801 (2003)

Hagley, E.W., Deng, L., Kozuma, M., Wen, J., Helmerson, K., Rolston, S.L., Phillips, W.D.: A well-collimated quasi-continuous atom laser. Science **283**, 1706 (1999)

He, P.R., Zhang, W.J., Li, Q.: Some further development on the eigensystem approach for graph isomorphism detection. J. Frankl. Inst. **342**, 657 (2005)

Heckenberg, N.R., McDuff, R., Smith, C.P., White, A.G.: Generation of optical phase singularities by computer-generated holograms. Opt. Lett. **17**, 221 (1992)

Hillery, M., Bergou, J., Feldman, E.: Quantum walks based on an interferometric analogy. Phys. Rev. A **68**, 032314 (2003)

Hines, A.P., Stamp, P.C.E.: Quantum walks, quantum gates, and quantum computers. Phys. Rev. A **75**, 062321 (2007)

Hoshino, S., Ichida, K.: Solution of partial differential equations by a modified random walk. Numer. Math. **18**, 61 (1971)

Jaksch, D.: Optical lattices, ultracold atoms and quantum information processing. Contemp. Phys. **45**, 367 (2004)

Jeong, H., Paternostro, M., Kim, M.S.: Simulation of quantum random walks using the interference of a classical field. Phys. Rev. A **69**, 012310 (2004)

Jessen, P.S., Deutsch, I.H.: In: Bederson, B., Walther, H. (eds.) Advances in Atomic and Molecular Physics, vol. 37, p. 95. Academic Press, Cambridge (1996)

Joo, J., Knight, P.L., Pachos, J.K.: Single atom quantum walk with 1D optical superlattices. J. Mod. Opt. **54**, 1627 (2007)

Karski, M., Förster, L., Choi, J.-M., Steffen, A., Alt, W., Meschede, D., Widera, A.: Quantum walk in position space with single optically trapped atoms. Science **325**, 174 (2009)

Kay, A., Pachos, J.K.: Quantum computation in optical lattices via global laser addressing. New J. Phys. **6**, 126 (2004)

Keating, J.P., Linden, N., Matthews, J.C.F., Winter, A.: Localization and its consequences for quantum walk algorithms and quantum communication. Phys. Rev. A **76**, 012315 (2007)

Kempe, J.: RANDOM'03: Proceedings of the 7th International Workshop on Randomization and Approximation Techniques in Computer Science. Lecture Notes in Computer Science, vol. 2764, pp. 354–369. Springer (2003a). arXiv:quant-ph/0205083

Kempe, J.: Quantum random walks: an introductory overview. Contemp. Phys. **44**, 307 (2003b)

Kendon, V.: Quantum walks on general graphs. Int. J. Quantum Inf. **4**, 791 (2006a)

Kendon, V.M.: A random walk approach to quantum algorithms. Philos. Trans. R. Soc. A **364**, 3407 (2006b)

Kendon, V.: Decoherence in quantum walks – a review. Math. Struct. Comput. Sci. **17**, 1169 (2007)

Kendon, V., Sanders, B.C.: Complementarity and quantum walks. Phys. Rev. A **71**, 022307 (2005)

Kendon, V., Tregenna, B.: QCMC'02: Proceedings of the 6th International Conference on Quantum Communication, Measurement and Computing, p. 463. Rinton Press, (2002). arXiv:quant-ph/0210047

Kendon, V., Tregenna, B.: Decoherence can be useful in quantum walks. Phys. Rev. A **67**, 042315 (2003)

Kendona, V., Maloyera, O.: Optimal computation with non-unitary quantum walks. Theor. Comput. Sci. **394**, 187 (2008)

Kilian, L., Taylor, M.P.: Why is it so difficult to beat the random walk forecast of exchange rates? J. Int. Econ. **60**, 85 (2003)

Kim, J., Benson, O., Kan, H., Yamamoto, Y.: A single-photon turnstile device. Nature **397**, 500 (1999)

Kim, M.S., Son, W., Bužek, V., Knight, P.L.: Entanglement by a beam splitter: nonclassicality as a prerequisite for entanglement. Phys. Rev. A **65**, 032323 (2002)

Kis, Z., Renzon, F.: Qubit rotation by stimulated Raman adiabatic passage. Phys. Rev. A **65**, 032318 (2002)

Knight, P.L., Roldán, E., Sipe, J.E.: Optical cavity implementations of the quantum walk. Opt. Commun. **227**, 147 (2003a)

Knight, P.L., Roldán, E., Sipe, J.E.: Quantum walk on the line as an interference phenomenon. Phys. Rev. A **68**, 020301 (2003b)

Köbler, J.: On graph isomorphism for restricted graph classes. In: Logical Approaches to Computational Barriers. Lecture Notes in Computer Science, vol. 3988, pp. 241–256. Springer, Berlin (2006)

Konno, N.: Limit theorem for continuous-time quantum walk on the line. Phys. Rev. E **72**, 26113 (2005)

Konno, N., Namiki, T., Soshi, T.: Symmetry of distribution for the one-dimensional Hadamard walk. Interdiscip. Inf. Sci. **10**, 11 (2004)

Košík, J., Bužek, V.: Scattering model for quantum random walks on a hypercube. Phys. Rev. A **71**, 012306 (2005)

Košík, J., Bužek, V., Hillery, M.: Quantum walks with random phase shifts. Phys. Rev. A **74**, 022310 (2006)

Krovi, H., Brun, T.A.: Hitting time for quantum walks on the hypercube. Phys. Rev. A **73**, 032341 (2006a)

Krovi, H., Brun, T.A.: Quantum walks with infinite hitting times. Phys. Rev. A **74**, 042334 (2006b)

Krovi, H., Brun, T.A.: Quantum walks on quotient graphs. Phys. Rev. A **75**, 062332 (2007)

Kurtsiefer, C., Mayer, S., Zarda, P., Weinfurter, H.: Stable solid-state source of single photons. Phys. Rev. Lett. **85**, 290 (2000)

Kwiat, P.G., Mattle, K., Weinfurter, H., Zeilinger, A.: New high-intensity source of polarization-entangled photon pairs. Phys. Rev. Lett. **75**, 4337 (1995)

Leach, J., Padgett, M.J., Barnett, S.M., Franke-Arnold, S., Courtial, J.: Measuring the orbital angular momentum of a single photon. Phys. Rev. Lett. **88**, 257901 (2002)

Leach, J., Courtial, J., Skeldon, K., Barnett, S.M.: Interferometric methods to measure orbital and spin, or the total angular momentum of a single photon. Phys. Rev. Lett. **92**, 013601 (2004)

Lee, P.J., Anderlini, M., Brown, B.L., Sebby-Strabley, J., Phillips, W.D., Porto, J.: Sublattice addressing and spin-dependent motion of atoms in a double-well lattice. Phys. Rev. Lett. **99**, 020402 (2007)

Li, Y., Hang, C., Ma, L., Zhang, W., Huang, G.: Quantum random walks in a coherent atomic system via electromagnetically induced transparency. J. Opt. Soc. Am. B **25**, C39 (2008)

Loke, T., Wang, J.B.: An efficient quantum circuit analyser on qubits and qudits. Comput. Phys. Commun. **182**, 2285 (2011)

López, C.C., Paz, J.P.: Phase-space approach to the study of decoherence in quantum walks. Phys. Rev. A **68**, 052305 (2003)

Lounis, B., Moerner, W.E.: Single photons on demand from a singlemolecule at room temperature. Nature **407**, 491 (2000)

Luo, X., Zhu, X., Wu, Y., Feng, M., Gao, K.: All-quantized Jaynes-Cummings interaction for a trapped ultracold ion. Phys. Lett. A **237**, 354 (1998)

Maloyer, O., Kendon, V.: Decoherence versus entanglement in coined quantum walks. New J. Phys. **9**, 87 (2007)

Mandel, O., Greiner, M., Widera, A., Rom, T., Hänsch, T.W., Bloch, I.: Coherent transport of neutral atoms in spin-dependent optical lattice potentials. Phys. Rev. Lett. **91**, 010407 (2003a)

Mandel, O., Greiner, M., Widera, A., Rom, T., Hänsch, T.W., Bloch, I.: Controlled collisions for multiparticle entanglement of optically trapped atoms. Nature **425**, 937 (2003b)

Manouchehri, K., Wang, J.B.: Continuous-time quantum random walks require discrete space. J. Phys. A **40**, 13773 (2007)

Manouchehri, K., Wang, J.B.: Quantum walks in an array of quantum dots. J. Phys. A **41**, 065304 (2008a)

Manouchehri, K., Wang, J.B.: Solid State implementation of quantum random walks on general graphs. In: Solid-State Quantum Computing: Proceedings of the 2nd International Workshop on Solid-State Quantum Computing & Mini-School on Quantum Information Science, Taipei. AIP Conference Proceedings, vol. 1074, pp. 56–61. Springer (2008b)

Manouchehri, K., Wang, J.B.: Quantum random walks without walking. Phys. Rev. A **80**, 060304(R) (2009)

Marquezino, F.L., Portugal, R.: Mixing times in quantum walks on the hypercube. Phys. Rev. A **77**, 042312 (2008)

Metcalf, H.J., der Straten, P.V.: Laser Cooling and Trapping. Springer, New York (1999a). Chapter 1

Metcalf, H.J., der Straten, P.V.: Laser Cooling and Trapping. Springer, New York (1999b). Chapter 8

Meyer, D.A.: From quantum cellular automata to quantum lattice gases. J. Stat. Phys. **85**, 551 (1996)

Mompart, J., Eckert, K., Ertmer, W., Birkl, G., Lewenstein, M.: Quantum computing with spatially delocalized qubits. Phys. Rev. Lett. **90**, 147901 (2003)

Monroe, C., Meekhof, D.M., King, B.E., Jefferts, S.R., Itano, W.M., Wineland, D.J.: Resolved-sideband Raman cooling of a bound atom to the 3D zero-point energy. Phys. Rev. Lett. **75**, 4011 (1995)

Monroe, C., Meekhof, D.M., King, B.E., Wineland, D.J.: A "Schrödinger Cat" Superposition State of an Atom. Science **272**, 1131 (1996)

Montanaro, A.: Quantum walks on directed graphs. Quantum Inf. Comput. **7**, 93 (2007)

Moore, C., Russell, A.: RANDOM'02: Proceedings of the 6th International Workshop Randomization and Approximation in Computer Science. Lecture Notes in Computer Science, vol. 2483, pp. 164–178. Springer (2002). arXiv:quant-ph/0104137

Morsch, O., Oberthaler, M.: Dynamics of Bose-Einstein condensates in optical lattices. Rev. Mod. Phys. **78**, 179 (2006)

Mosley, P.J., Croke, S., Walmsley, I.A., Barnett, S.M.: Experimental realization of maximum confidence quantum state discrimination for the extraction of quantum information. Phys. Rev. Lett. **97**, 193601 (2006)

Moya-Cessa, H., Bužek, V., Knight, P.: Power broadening and shifts of micromaser lineshapes. Opt. Commun. **85**, 267 (1991)

Mülken, O., Blumen, A.: Continuous-time quantum walks: models for coherent transport on complex networks. Phys. Rep. **502**, 37 (2011)

Nakahara, M., Ohmi, T.: Quantum Computing: From Linear Algebra to Physical Realizations. CRC Press, Taylor and Francis Group, Boca Raton (2008). Chapter 14

Nayak, A., Vishwan, A.: Quantum Walk on the Line. DIMACS Technical Report **43** (2000). arXiv:quant-ph/0010117

Oberthaler, M.: Lecture Notes: Matter Waves and Light, Technical Report, Universität Heidelberg, Kirchhoff-Institut für Physik (2007)

Oliveira, A.C., Portugal, R., Donangelo, R.: Decoherence in two-dimensional quantum walks. Phys. Rev. A **74**, 012312 (2006)

Omar, Y., Paunković, N., Bose, L.S.S.: Quantum walk on a line with two entangled particles. Phys. Rev. A **74**, 042304 (2006)

Pandey, D., Satapathy, N., Meena, M.S., Ramachandran, H.: Quantum walk of light in frequency space and its controlled dephasing. Phys. Rev. A **84**, 042322 (2011)

Paris, M.G.: Displacement operator by beam splitter. Phys. Lett. A **217**, 78 (1996)

Paris, M.G.: Joint generation of identical squeezed states. Phys. Lett. A **225**, 28 (1997)

Patel, A., Raghunathan, K.S., Rungta, P.: Quantum random walks do not need a coin toss. Phys. Rev. A **71**, 032347 (2005)

Pathak, P.K., Agarwal, G.S.: Quantum random walk of two photons in separable and entangled states. Phys. Rev. A **75**, 032351 (2007)

Peil, S., Porto, J.V., Tolra, B.L., Obrecht, J.M., King, B.E., Subbotin, M., Rolston, S.L., Phillips, W.D.: Patterned loading of a Bose-Einstein condensate into an optical lattice. Phys. Rev. A **67**, 051603 (2003)

Peruzzo, A., Lobino, M., Matthews, J.C.F., Matsuda, N., Politi, A., Poulios, K., Zhou, X.-Q., Lahini, Y., Ismail, N., Wörhoff, K., Bromberg, Y., Silberberg, Y., et al.: Quantum walks of correlated photons. Science **329**, 1500 (2010)

Pioro-Ladrière, M., Abolfath, M.R., Zawadzki, P., Lapointe, J., Studenikin, S.A., Sachrajda, A.S., Hawrylak, P.: Charge sensing of an artificial $H_2^+$ molecule in lateral quantum dots. Phys. Rev. B **72**, 125307 (2005)

Price, M.D., Somaroo, S.S., Dunlop, A.E., Havel, T.F., Cory, D.G.: Generalized methods for the development of quantum logic gates for an NMR quantum information processor. Phys. Rev. A **60**, 2777 (1999)

Raimond, J.M., Brune, M., Haroche, S.: Colloquium: manipulating quantum entanglement with atoms and photons in a cavity. Rev. Mod. Phys. **73**, 565 (2001)

Regensburger, A., Bersch, C., Hinrichs, B., Onishchukov, G., Schreiber, A.: Phys. Rev. Lett. **107**, 233902 (2011)

Reitzner, D., Hillery, M., Feldman, E., Bužek, V.: (2009). `arXiv:quant-ph/0805.1237v3`

Ribeiro, P., Milman, P., Mosseri, R.: Aperiodic quantum random walks. Phys. Rev. Lett. **93**, 190503 (2004)

Richter, P.C.: Almost uniform sampling via quantum walks. New J. Phys. **9**, 72 (2007a)

Richter, P.C.: Quantum speedup of classical mixing processes. Phys. Rev. A **76**, 042306 (2007b)

Rohde, P.P., Schreiber, A., Štefaňák, M., Jex, I., Silberhorn, C.: Multi-walker discrete time quantum walks on arbitrary graphs, their properties and their photonic implementation. New J. Phys. **13**, 013001 (2011)

Roldán, E., Soriano, J.C.: Optical implementability of the two-dimensional quantum walk. J. Mod. Opt. **52**, 2649 (2005)

Romanelli, A., Siri, R., Abal, G., Auyuanet, A., Donangelo, R.: Decoherence in the quantum walk on the line. Physica A **347**, 137 (2005)

Ryan, C.A., Laforest, M., Boileau, J.C., Laflamme, R.: Experimental implementation of a discrete-time quantum random walk on an NMR quantum-information processor. Phys. Rev. A **72**, 062317 (2005)

Sanders, B.C., Bartlett, S.D.: Quantum quincunx in cavity quantum electrodynamics. Phys. Rev. A **67**, 042305 (2003)

Sansoni, L., Sciarrino, F., Vallone, G., Mataloni, P., Crespi, A., Ramponi, R., Osellame, R.: Two-particle Bosonic-Fermionic quantumwalk via integrated photonics. Phys. Rev. Lett. **108**, 010502 (2012)

Schmitz, H., Matjeschk, R., Schneider, C., Glueckert, J., Enderlein, M., Huber, T., Schaetz, T.: Quantum walk of a trapped ion in phase space. Phys. Rev. Lett. **103**, 090504 (2009)

Schöll, J., Schöll-Paschingerb, E.: Classification by restricted random walks. Pattern Recognit. **36**, 1279 (2003)

Schreiber, A., Cassemiro, K.N., Potoček, V., Gábris, A., Mosley, P.J., Andersson, E., Jex, I., Silberhorn, C.: Photons walking the line: a quantum walk with adjustable coin operations. Phys. Rev. Lett. **104**, 050502 (2010)

Schreiber, A., Cassemiro, K.N., Potoček, V., Gábris, A., Jex, I., Silberhorn, C.: Photons walking the line: a quantum walk with adjustable coin operations. Phys. Rev. Lett. **106**, 180403 (2011)

Schreiber, A., Gábris, A., Rohde, P.P., Laiho, K., Štefanák, M., Potocek, V., Hamilton, C., Jex, I., Silberhorn, C.: A 2D quantum walk simulation of two-particle dynamics. Science **336**, 55 (2012)

Sessionsa, R.B., Orama, M., Szczelkuna, M.D., Halforda, S.E.: Random walk models for DNA synapsis by resolvase. J. Mol. Biol. **270**, 413 (1997)

Shapira, D., Biham, O., Bracken, A.J., Hackett, M.: One-dimensional quantum walk with unitary noise. Phys. Rev. A **68**, 062315 (2003)

Shenvi, N., Kempe, J., Whaley, K.B.: Quantum random-walk search algorithm. Phys. Rev. A **67**, 052307 (2003)

Shiau, S.Y., Joynt, R., Coppersmith, S.N.: Physically-motivated dynamical algorithms for the graph isomorphism problem. Quantum Inf. Comput. **5**, 492 (2005)

Shore, B.W.: The Theory of Coherent Atomic Excitation. Wiley-Interscience, New York (1990a). Chapter 3

Shore, B.W.: The Theory of Coherent Atomic Excitation. Wiley-Interscience, New York (1990b) Chapter 2

Shore, B.W.: The Theory of Coherent Atomic Excitation. Wiley-Interscience, New York (1990c). Chapter 13

Shore, B.W., Knight, P.L.: Topical review: the Jaynes-Cummings model. J. Mod. Opt. **40**, 1195 (1993)

Slater, J.C.: Atomic radii in crystals. J. Chem. Phys. **41**, 3199 (1964)

Solenov, D., Fedichkin, L.: Continuous-time quantum walks on a cycle graph. Phys. Rev. A **73**, 012313 (2006a)

Solenov, D., Fedichkin, L.: Nonunitary quantum walks on hypercycles. Phys. Rev. A **73**, 012308 (2006b)

Spreeuw, R.J.C.: A classical analogy of entanglement. Found. Phys. **28**, 361 (1998)

Spreeuw, R.J.C.: Classical wave-optics analogy of quantum-information processing. Phys. Rev. A **63**, 062302 (2001)

Stefanak, M., Kiss, T., Jex, I., Mohring, B.: The meeting problem in the quantum walk. J. Phys. A **39**, 14965 (2006)

Stewart, I.: Where drunkards hang out. Nature **413**, 686 (2001)

Strauch, F.W.: Connecting the discrete- and continuous-time quantum walks. Phys. Rev. A **74**, 030301 (2006a)

Strauch, F.W.: Relativistic quantum walks. Phys. Rev. A **73**, 054302 (2006b)

Sutton, B.D.: Computing the complete CS decomposition. Computing the complete CS decomposition. Numer. Algorithms **50**, 33–65 (2009)

Townsend, J.S.: A Modern Approach to Quantum Mechanics. University Science Books, Sausalito (2000a)

Townsend, J.S.: A Modern Approach to Quantum Mechanics. University Science Books, Sausalito (2000b). Chapter 14

Trautt, Z.T., Upmanyu, M., Karma, A.: Interface mobility from interface random walk. Science **314**, 632 (2006)

Travaglione, B.C., Milburn, G.J.: Implementing the quantum random walk. Phys. Rev. A **65**, 032310 (2002)

Tregenna, B., Flanagan, W., Maile, R., Kendon, V.: Controlling discrete quantum walks: coins and initial states. New J. Phys. **5**, 83 (2003)

van den Engh, G., Sachs, R., Trask, B.: Estimating genomic distance from DNA sequence location in cell nuclei by a random walk model. Science **257**, 1410 (1992)

Vaziri, A., Weihs, G., Zeilinger, A.: Superpositions of the orbital angular momentum for applications in quantum experiments. J. Opt. B **4**, S47 (2002)

Venegas-Andraca, S.E.: (2012). arXiv:1201.4780

Venegas-Andraca, S.E., Bose, S.: (2009). arXiv:0901.3946

Venegas-Andraca, S.E., Ball, J.L., Burnett, K., Bose, S.: Quantum walks with entangled coins. New J. Phys. **7**, 221 (2005)

Wang, J.B., Midgley, S.: Quantum waveguide theory: a direct solution to the time-dependent Schrödinger equation. Phys. Rev. B **60**, 13668 (1999)

Watrous, J.: J. Comput. Syst. Sci. **62**, 376 (2001)

Wineland, D.J., Barrett, M., Britton, J., Chiaverini, J., DeMarco, B., Itano, W.M., Jelenkovic, B., Langer, C., Leibfried, D., Meyer, V., Rosenband, T., Schätz, T.: Quantum information processing with trapped ions. Philos. Trans. R. Soc. Lond. **361**, 1349 (2003)

Wódkiewicz, K., Eberly, J.H.: Coherent states, squeezed fluctuations, and the SU(2) and SU(1, 1) groups in quantum-optics applications. J. Am. Opt. Soc. B **2**, 458 (1985)

Wójcik, A., Łuczak, T., Kurzyński, P., Grudka, A., Bednarska, M.: Quasiperiodic dynamics of a quantum walk on the line. Phys. Rev. Lett. **93**, 180601 (2004)

Wright, K.C., Leslie, L.S., Bigelow, N.P.: Optical control of the internal and external angular momentum of a Bose-Einstein condensate. Phys. Rev. A **77**, 041601 (2008)

Wu, Y., Deng, L.: Ultraslow optical solitons in a cold four-state medium. Phys. Rev. Lett. **93**, 143904 (2004)

Würtz, P., Langen, T., Gericke, T., Koglbauer, A., Ott, H.: (2009). arXiv:0903.4837v1

Xiang-bin, W.: Theorem for the beam-splitter entangler. Phys. Rev. A **66**, 024303 (2002)

Xu, X.-P.: Continuous-time quantum walks on one-dimensional regular networks. Phys. Rev. E **77**, 061127 (2008)

Xue, P., Sanders, B.C., Leibfried, D.: Quantum walk on a line for a trapped ion. Phys. Rev. Lett. **103**, 183602 (2009)

Yin, J.: Realization and research of optically-trapped quantum degenerate gases. Phys. Rep. **430**, 1 (2006)

Yin, Y., Katsanos, D.E., Evangelou, S.N.: Quantum walks on a random environment. Phys. Rev. A **77**, 022302 (2008)

Yurke, B., McCall, S.L., Klauder, J.R.: SU(2) and SU(1, 1) interferometers. Phys. Rev. A **33**, 4033 (1986)

Zähringer, F., Kirchmair, G., Gerritsma, R., Solano, E., Blatt, R., Roos, C.F.: Realization of a quantum walk with one and two trapped ions. Phys. Rev. Lett. **104**, 100503 (2010)

Zhang, P., Ren, X.-F., Zou, X.-B., Liu, B.-H., Huang, Y.-F., and Guo, G.-C.: Demonstration of one-dimensional quantum random walks using orbital angular momentum of photons. Phys. Rev. A **75**, 052310 (2007)

Zhang, P., Liu, B.-H., Liu, R.-F., Li, H.-R., Li, F.-L., Guo, G.-C.: Implementation of one-dimensional quantum walks on spin-orbital angular momentum space of photons. Phys. Rev. A **81**, 052322 (2010)

Zhao, Z., Du, J., Li, H., Yang, T., Chen, Z.-B., Pan, J.-W.: (2002). arXiv:quant-ph/0212149

Zou, X., Dong, Y., Guo, G.: Optical implementation of one-dimensional quantum random walks using orbital angular momentum of a single photon. New J. Phys. **8**, 81 (2006)

# Index

## A

a.c. Stark effect, 180–181
Anti-Stokes radiation, 182
Atom-field interaction
  dipole approximation, 168–170
  energy, 165–167
  multipole expansion, 167–168
Atomic coordinate system, 165–166
Atom optics
  atom-field interaction
    dipole approximation, 168–170
    energy, 165–167
    multipole expansion, 167–168
  single-photon interaction
    a.c. Stark effect, 180–181
    interaction picture, 172–173
    Jaynes-Cummings model, 177–179
    Rabi model, 170–172
    RWA, 173–175
    separable Hamiltonians, 176
  two-photon interaction
    adiabatic elimination, 184–185
    Raman process, 181–184
    STIRAP, 185–187
Average mixing, 25

## B

Beam splitters
  Baker-Campbell-Hausdorff operator, 197
  dielectric medium, 196
  evolution operators, 197
  optical mode mixer, 196
  transmittance and reflectance coefficients, 198
Bloch rotations
  non-resonance Raman process, 194

Rabi flopping, 192–193
sphere
  Cartesian coordinates, 190–191
  orthogonal axes, 191–192
  schematic diagram, 189–190
STIRAP, 192

## C

Cavity QED
  action of evolution operator, 83
  coherent state Wigner function, 86
  conditional translation operator, 87–88
  Gaussian distribution, 84–85
  periodic sequence, Raman pulses, 85
  Rydberg atom injection, 81–82
  superposition state, 83–84
  two-level atom, 87
Classical radiation field
  electric component, 154
  Jones vector, 155
  Maxwell's equations, 153
Classical random walks. *See* Random walks
Coined quantum walks. *See* Discrete-time
    quantum walks
Continuous-time quantum walks
  characteristic probability distribution,
    20–21
  characteristic signature-Gaussian transition,
    22, 24
  decision tree, 19
  *vs.* discrete-time, 19–20
  nodes, 20–21
  orthogonality of position states, 22
  ring shaped array, 23
  transition rate matrix, 19–20
  types of networks, 24

J. Wang and K. Manouchehri, *Physical Implementation of Quantum Walks*,
Quantum Science and Technology, DOI 10.1007/978-3-642-36014-5,
© Springer-Verlag Berlin Heidelberg 2014